GENOMICS

GENOMICS

How Genome Sequencing Will Change Our Lives

Rachael Pells

Cornerstone Press

1 3 5 7 9 10 8 6 4 2

Cornerstone Press
20 Vauxhall Bridge Road
London SW1V 2SA

Cornerstone Press is part of the Penguin Random House
group of companies whose addresses can be found at
global.penguinrandomhouse.com

Illustrations on pp. 17 and 30 based on drawings by Anthony Johnson

First published in the UK by Cornerstone Press in 2022

www.penguin.co.uk

A CIP catalogue record for this book is available from the British Library

ISBN 9781847943408

Typeset in 9.5/18 pt Exchange Text
by Integra Software Services Pvt. Ltd, Pondicherry

Printed and bound in Great Britain by Clays Ltd, Elcograf S.p.A.

The authorised representative in the EEA is Penguin Random House
Ireland, Morrison Chambers, 32 Nassau Street, Dublin D02 YH68

Penguin Random House is committed to a sustainable future for
our business, our readers and our planet. This book is made from
Forest Stewardship Council® certified paper.

Contents

Introduction

We humans are fascinated with ourselves. What causes the similarities between us, and what makes us different? Why is it that some people live long and healthy lives, and others don't? For centuries, the brightest minds have sought to determine what it is that makes us who we are, but also how we might live better, stronger and longer.

Genomics – that is, the study of all of our thousands of genes – is finally offering some answers to some of the biggest, longest-held questions about humanity, as well as the world we live in. DNA studies were once long and mystifying processes, but groundbreaking sequencing technologies now mean that, at the press of a button, scientists are able to create a map of all the genes we contain – and deduce a lot of the key messages those genes send.

This map is called a genome, and it contains many important clues about us, from our ancestry to the way our bodies respond to diseases, medication and ageing. In fact, genomics can be applied to just about every part of who we are and what we do – from the different traits we inherit and our resistance to drugs to humanity's ability to solve crimes and reduce the impacts of climate change. The possibilities are vast and have the ability to transform the way we think about ourselves and the world around us.

At the same time, many of us would be willing to admit that we don't know much about genomics at all. That's understandable, considering the term itself didn't exist until the late 1980s, when an international moonshot initiative was launched to create a full blueprint of all human genes for the first time. Back then, the research was largely experimental and, as many point out, some of its practices wouldn't be considered ethical today. Researchers would take samples of their own blood to examine in the lab, for example, or draw straws among colleagues to see whose sample would be taken for that day's work. Today the practice is much better understood, regulated and shared – and the technologies to have been born out of it are used within an incredibly wide range of research disciplines.

It's entirely likely that one day soon, your genome will be sequenced by a small machine in your GP's office to help determine the best course of treatment for anything from antibiotics to drugs for anxiety and depression. Genome sequencing can also offer clues as to how to best modify treatments, and even tailor them to the individual – bringing us one big step closer to the dream of precision medicine.

And yet, as the saying goes, with great knowledge comes great responsibility – and there are many unknown paths left to go down in terms of ethics and the way we use genetic DNA going forward. For decades, popular culture as well as the media has warned of the dangers of creating 'designer babies' and playing God to potentially disastrous effect. Some of that has already become a reality: would-be parents can use genomics to check the DNA of embryos to see if they like what's there. The technology is only currently being used to screen against inheritable diseases, but it could be used in the future to select for intelligence and aesthetic traits. Scientists already have the tools they need to repair, modify or cut out unwanted genes entirely should they wish to - but the long-term impacts of this are unknown and experts haven't yet worked out what the rules of the game might be or how to play it.

As a society we are confused by the ethics of genomics. For example, we may express concerns over data protection, and yet at the same time send away for home-testing DNA kits, eager to share our most personal data to learn what traits we might carry in accordance with our genes. Many of us may well be appalled at the idea we might curate our children's looks or characteristics, pick one eye colour over another, even build a new biological hierarchy. And yet for some people, the ability to choose traits is a tempting prospect, something they believe would give their offspring the best chance of success. Whatever we think about the potential of genomics to help us, we have to be aware of the fact that, at its extreme, it can so easily merge with eugenics, the advocacy of controlled selective breeding beloved by twentieth-century fascist regimes and so rightly shunned.

Clearly, genomics, and specifically gene editing, is an ethical minefield that will only raise bigger and tougher questions as the research continues to evolve. What we choose to do with these tools as a society is up to us: but the first key to decision making is understanding. Rightly or wrongly, genomics will shape the future of our health, the future of our planet – our entire existence as a species – and so it's crucial we get it right.

1

How to map a life

When DNA was first identified in the late 1860s, scientists advanced one step further in their understanding of how the human body works, and even how it might be manipulated. But there was still a long way to go before theory could meet reality. In fact, experts didn't learn the full picture of how our DNA links together until around 140 years later, through a $2.7 billion (£2.2 billion) moonshot initiative called the Human Genome Project (HGP).

Arguably the most ambitious research project of modern times, the HGP was led by an international group of researchers with the aim of mapping the entire human genome – that is, a detailed reference for our complete genetic code. Launched on 1 October 1990, the project was initially funded by the US government, and later by the UK's Medical Research Council and the Wellcome Trust.

In order to help facilitate the research, in 1992 Wellcome set up the Sanger Institute, a dedicated laboratory on a factory-sized scale near Cambridge. Similar dedicated spaces were built in California and Texas, but sequencing took place at numerous universities and research centres across the UK, US, France, Germany, Japan and China.[1] By the time of the HGP's completion in 2003, at least eighteen countries had contributed in some form or another.

The reason such large labs and so many contributors were necessary was in part because of the huge scale of the project: there are 3.2 billion letters of DNA in the human body, and each one had to be identified and recorded. At a time when technology was much more limited, the project required all the expert eyes and ears possible and so researchers split up tasks evenly across labs, countries and continents. A crude map was created to allow contributors to each add in their respective pieces of the puzzle. But more than this, the HGP was designed to be collaborative, involving as many different laboratories and nationalities as possible to ensure the work was inclusive and reflective of the genetic expertise found all over the world. Moreover, it was important to both funders and scientific leaders that this was a global project, both to share the huge costs involved and to prevent any one party claiming ownership of the human genome – the secrets of our existence.

For the many, many researchers involved in mapping the first human genome, the project itself presented an opportunity to be part of history. 'It was the complete opposite of the lingering old-fashioned belief that science was for the alpha male, the brilliant lone genius, all that rubbish,' says Julian Parkhill, who joined the Sanger Institute in 1997 and eventually became its head of pathogen genomics. 'This was a very large community of scientists all working together towards a common goal.'

Stephan Beck, a leading medical genomicist, was head of human sequencing at the Sanger Institute for some of that time, where he played a leading role in the sequencing and analysis of human and mouse genomes. He remembers the excitement of the time well: 'We knew it was going to be a unique project – no species has ever been capable of sequencing its own genetic content before and this was something that was only going to happen once,' he says. Multiple donors were required throughout the project in order for researchers to collect enough human DNA to work with (using blood and sperm samples), but their identities were kept anonymous. 'In my experience, there was never any shortage of volunteers,' says Beck.

But with so much at stake, tension was growing between the public and private research sectors, each vying to win the prize of being the first named group to solve the puzzle of human existence. By the late 1990s an American biotechnologist and entrepreneur named Craig Venter was making his frustrations over the slow pace of public projects known. Formerly an employee at the US National Institutes of Health (NIH), Venter had pioneered new techniques in genetics research which were quicker than those being used by the HGP partners, and he saw a business opportunity in the mapping of the human genome. Confident that he could do better, Venter sought funding from the private sector and went about hiring his own research team to map the human genome himself.[2] 'Government-funded efforts were painstakingly slow,' says Parkhill. Meanwhile, 'here was this private industry individual saying, "I've got some money, I'm going to do it all myself." And he tried to persuade the US government to let him do it instead.'

Venter's intervention became a sore point in the research community for several reasons. For one thing, the HGP's public networks had been established on the condition of openness and transparency. This was formalised in a meeting in Bermuda in February 1996, when the HGP partners agreed to share their progress on public research databases every twenty-four hours. The move towards open data made researchers accountable for their work and helped to ensure that work was not accidentally repeated across borders. It allowed for a more collaborative approach – research teams could cross-check data and point out any potential errors – but it also adhered to the principles of making publicly funded science openly available for the public good, rather than hidden away behind paywalls or lost in private storage drives.

Venter declined to be part of the arrangement, initially keeping his data locked away from other contributors. He also accused the US government of wasting public funding by making the operation so far-reaching, involving 'armies of scientists' without any plan in place to make the money back through innovation. Venter's plan, by contrast, was to patent the genes he mapped out and sell access to the data through a subscription service.[3] Publicly funded researchers taking part in the open approach to data questioned Venter's motives and lack of transparency.[4]

Ultimately, Venter's mission didn't result in the public efforts being shut down as some might have feared. If anything his private company, Celera, helped to speed up the public effort by adding a huge amount of pressure on governments to get the job done first. With the HGP facing the prospect of losing its claim to the discovery of the century, the Wellcome Trust pledged to double its investment in the public efforts, turning the project into a race.[5] 'The whole point of the Human Genome Project was for it to be a joint international venture that kept the findings in the public domain,' says Parkhill. 'If Venter had managed to persuade the US government not to fund this as a public endeavour ... we would have had a private human genome, which we would have had to pay for access to.'

On 25 June 2000, US and UK leaders made a joint statement, joined by both Venter and NIH director Francis Collins, to announce that a draft genome was complete. There were handshakes all round, the consensus being that the different groups involved had reached the milestone together, that it was a tie. The Human Genome Project was expected to take around fifteen years to complete, but researchers had cracked the code more than five years early. There's no doubt Venter's company contributed significantly to the final human genome map. But bitter arguments still linger over who really got there first – whether Celera could have achieved what it did without using the public data that was being published in real time, and vice versa.

Ultimately, the data published that day all was overtaken anyway. More accurate maps of the human genome were published later on, until the HGP was officially closed in 2003 with publication of a final, full human genome, free for all to access online. The Human Genome Project has been likened to an Apollo space mission for biology, offering more detailed clues to our existence than many could even have hoped for. Not only did the project confirm some of the genetic links scientists expected from DNA, such as eye and hair colour, it also revealed some brand new and unexpected discoveries about our ancestry – the fact that ancient humans mated with Neanderthals, for example.

DNA studies that have followed since have drawn on the project's findings to determine other important insights to human health, too, for example by pinpointing mutated genes that can lead to cancer.

Others have helped to develop new drugs for conditions such as cystic fibrosis and asthma. In other words, the importance of the project to modern medicine cannot be overstated.

Understanding genomics

Where once the sequencing tools used to determine DNA patterns were expensive and time-consuming, these days the technology is fast, efficient and relatively cheap compared to other medical tools. This means that we are continuing to learn new things about the human genome at an increasingly fast pace. Search on a research publication database today for papers with 'genomics' in the title and you'll be presented with thousands of exciting findings published by researchers across different disciplines in labs all over the world,[6] with the number increasing every week. More exciting still are the possibilities now presented by genome editing for improving human health, which I will explore later.

But first, to understand more about how genomics works, it helps to start with the basics. You'll probably already know that our bodies are made up of cells, which contain DNA. These famous double-helix strands are made up of four components, or DNA 'bases', known as A, C, G and T. The four letters stand for the chemicals adenine (A), cytosine (C), guanine (G) and thymine (T). These molecules couple up: G with C and T with A. You don't really need to know why or how, but it's useful to know that they do, and that these 'base pairs' become the rungs of the recognisable DNA ladder, the double helix. Together, these DNA bases form a code, or 'sequence', which effectively spells out the millions of different instructions – known as genes – needed for our bodies to function. Genes tell our cells how to make structural proteins and enzymes – the building blocks of the body that all our tissues and organs are made of. There are approximately 20,500 genes in the human body, and together they make a genome.

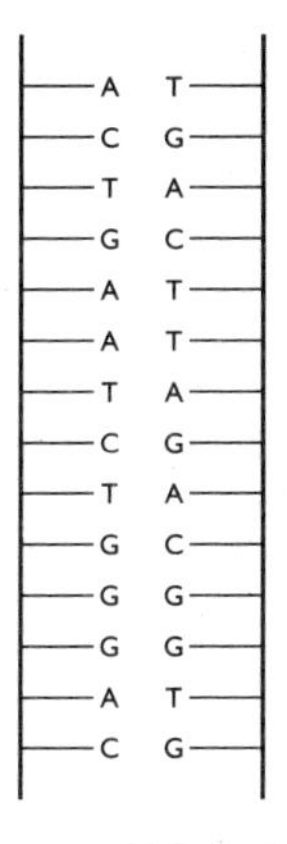

The DNA ladder, laid out as if flat, showing the base pairs

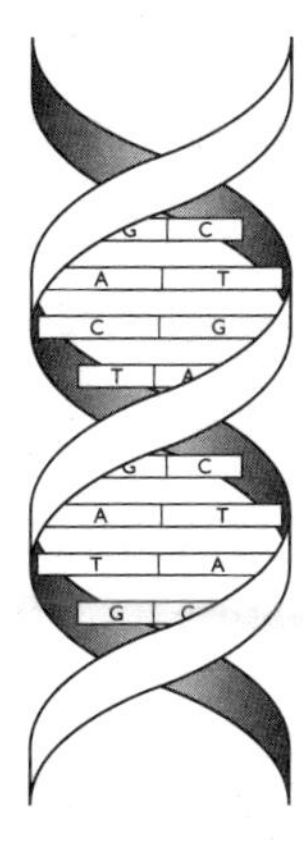

A three-dimensional model of the DNA double helix, with base pairs

A genome is essentially a map of all your DNA. Only about 1.5 per cent of it is made up of genes, however. A 10 per cent chunk of the genome is regulatory information and the rest is considered to be nonsense coding, which we've inherited over millions of years of evolution. Every living thing has its own genome, and each one contains the information needed to build and maintain an organism throughout its life. The human genome contains around 3.2 billion letters of DNA, which are split into twenty-three threads called chromosomes. If it were possible to unravel all your DNA out into a long, thin line, it would reach the equivalent of nearly seventy trips from the Earth to the Sun and back – the sheer scale of it is almost impossible to imagine.

The map created by the Human Genome Project provided a starting point for scientists to reference; now experts are trying to interpret the map to find out what it means in practice. New technologies are allowing scientists to interpret more of that coding by the day to help us understand why and how our bodies can go wrong, or how our quality of life might be improved. To get a fuller picture of a DNA strand, scientists can read its structure using specially designed equipment, either in their own labs or by sending samples away to external companies, in a process called 'sequencing'.

Advances in genetic sequencing technologies

Sequencing DNA effectively means determining the specific ordering of the A, C, G and T chemical building blocks it contains in order to establish the exact piece of code. The technologies used for this process have really come along since the early days of the HGP, when mapping out the first human genome was a laborious process. In the 1990s, researchers sequenced with pencil and paper for a time, using a system invented by the biochemist Fred Sanger, who became the first person to sequence the full DNA genome of anything – a virus of *E. coli* called phi

X 174 – and later won the Nobel Prize in Chemistry for his discovery. These first sequencing methods were a little like early analogue photography: researchers would label different DNA base pairs with a radioactive formula and leave it on a piece of film overnight. The next morning, they developed the film, and the recognisable DNA ladder was visible. Each piece of DNA had to be separately isolated and amplified by hand.

The Human Genome Project was made possible through the automation of that process. Just as important as the task of mapping out the human genome was the need to develop the technology required to understand and analyse it along the way. 'It was absolutely clear when the HGP first started, that the technology did not yet exist to complete it,' recalls Parkhill. 'There was an assumption that the technology would improve and get better along the way, and that's something that's been true of sequencing all along.'

According to Sharon Peacock, a leading microbiologist and executive director of the Covid-19 Genomics UK

Consortium (COG-UK), the first genome of a bacterium like *E. coli* would have taken eighteen months to sequence back in 2001. 'Twelve years later, scientists were doing it in a day,' she says. Not only this, but the practice has become cheaper and more accessible. Back in 2001, the cost of sequencing one *E. coli* genome sat at 'around half a million pounds; by 2013, it had plummeted to around £150'. The cost of sequencing the first human genome is often quoted at around $3 billion – around a dollar per gene. Today, anyone can pay to have their genome sequenced by online companies for as little as £300.

Costs have reduced dramatically for two reasons. Firstly, years ago a lot of the work to sequence DNA was done manually, which meant lots of people were involved and a lot of money was needed for their salaries. At the Sanger Institute, robots eventually replaced humans in the tedious tasks of extracting DNA and storing the data, and ultimately sequencing became much faster. Now, the entire process is automated: a blood sample is taken and the DNA is extracted by machine. The molecules are processed and the data is crunched by algorithms, not people. The entire process can be done in six to seven hours.

Secondly, as the technology becomes cheaper and more accessible, the materials required to sequence DNA are also becoming cheaper. For example, chemical companies can produce and sell the chemicals needed for the sequencing process at scale. The machinery is also becoming smaller – from big, noisy giants taking up a large amount of space in the early 1990s to coffee-machine-sized sequencers in the early 2000s.

At the time of writing, there are two primary technologies used for genetic sequencing. The first is sequencing by synthesis, most commonly used in a technique called Illumina fluorescent dye sequencing. Here, a sample of DNA is extracted from a piece of tissue, or a sample of blood or saliva, put on a glass slide and separated into two single-stranded DNA molecules using a chemical solution. These molecules are then exposed to something called a DNA polymerase enzyme which encourages them to replicate. The four DNA bases,

A, C, G and T, are given fluorescent dye markers, each in a different colour, to distinguish them. As each rung of the DNA ladder, or base pair, extends itself to replicate the next pieces of DNA in the code, a photo is taken to document the changes. This can then be used to examine the different colours, and therefore the genetic coding inside that sample.[7] Effectively, the researcher is using the tiny sample of DNA in order to re-form the rest of the genetic code – which the DNA does by itself, with some chemical encouragement. The molecules extend by one DNA base at a time, demonstrating what would happen next in the full sequence inside the body it came from.

This method is also known as short-read sequencing. Years ago, the method was slowly and painstakingly undertaken by hand, but Illumina technology accelerates the process by sequencing very large numbers of molecules in parallel, thereby allowing researchers to read a few hundred base pairs from many millions of molecules at the same time and therefore quickening the process.

The second, more recent method of genetic sequencing uses nanopore technology. This works by monitoring changes to an electric current, as a single, very long piece of DNA is threaded through an extremely small pore in a membrane. The resulting signal is decoded to provide the specific DNA or RNA sequence (RNA standing for 'ribonucleic acid', a molecule present in all living cells which controls how certain genes are expressed). Nanopore allows researchers to read tens of thousands of base pairs from a single molecule, which allows them to piece together the genome more efficiently, although in the past this has been less accurate than Illumina technology.

The two methods are used for different purposes, depending on the objectives of the study. Researchers who are interested in sequencing large numbers of samples for comparative purposes will tend to use Illumina because it is low in cost and high in accuracy, and therefore will define the small differences between samples very effectively. It also sequences in bulk, meaning the genomes need to be fairly big to ensure cost effectiveness. An example is the UK government's ambitious 100,000 Genomes Project, which was announced in 2012 with the goal of sequencing that number of genomes from National Health Service patients suffering from rare diseases. The project was undertaken using Illumina machines because speed was less of a factor (scientists didn't need the data back the same day) and the costs needed to be kept low because the project came from public funding. As the NHS moves towards human genome sequencing as a part of routine diagnostics, it will continue to use Illumina machines for similar reasons.

Nanopore devices are smaller and more portable and are usually directed to smaller projects. They are often used for sequencing out in the field – researchers working on the 2013 West African Ebola outbreak used them to test patient samples in order to track the development of the virus[8] – and where a fast turnaround is required for smaller numbers of samples with small genomes (fruit flies in a lab, for example). Most of the sequencing done by experts studying the spread of Covid-19 during the height of the pandemic was done locally using nanopore. Today, Oxford Nanopore Technologies sells a DNA sequencer small enough to attach to a mobile phone, opening up genomics to field work on the road.[9]

Sequencing in practice

With sequencing technologies now so cheap and available to labs across much of the world, the race is on to capture the genomes of different species of plants, animals and pathogens in order to better understand all life on Earth. Researchers across various subject fields have come together to take part in mass collecting drives such as the Darwin Tree of Life project, which aims to sequence the genomes of all 70,000 species of eukaryotic organisms (living things whose cells have a clearly defined nucleus, so not bacteria, for example) in Britain and Ireland.

The Earth BioGenome Project, launched in 2017, is a global initiative that aims to sequence and catalogue the genomes of all of Earth's eukaryotic species over a period of ten years. One of the initiative's co-founders, Harris Lewin, makes the case that twenty years after the first human genome was sequenced, less than half a per cent of species have had their genome sequenced – meaning the data has limited use in really furthering our understanding of the planet.

But when it comes to modern genome sequencing practices, there is a common saying among more cynically minded observers: that it's all really just a vanity project, or 'biological stamp collecting'. Why waste time and resources collecting more and more data like Pokémon with no direct purpose? 'It's a phrase that often gets thrown around, but I don't agree with it at all,' says Parkhill. 'There was a big push around the time the first human genome was published twenty years ago towards hypothesis-driven research – funders were concerned that too many people were just doing science without any particular direction. But the point is, when you're only doing hypothesis-driven research, it's very difficult to stumble upon novelty. You can find what you're looking for and prove it, but you can't find what you weren't looking for.'

One of the strengths of genomics, he argues, is that it's a 'completely agnostic way of looking at data' – researchers can approach the information with few preconceptions and explore big questions based on what they see. And indeed, plenty of useful discoveries have been made through serendipitous exploration of genetics, or blue-skies research – some medicines have been discovered through studies of things like snake venom, for example. All this supports the argument that, until researchers go out there and do their basic 'stamp collecting' they don't know what amazing things they might find. In other words, genetic sequencing holds great potential to enhance our knowledge of the planet and could bring tremendous benefits, perhaps especially in the areas we didn't think to look for them.

Introducing Crispr

Perhaps one of the biggest examples of serendipitous genetic discovery in our lifetime is the gene editing tool Crispr. By the turn of the millennium, gene editing technologies had progressed, but slowly. This was frustrating for researchers, who had worked out that it was possible to change a gene but didn't have the precision technology to do it safely.

In 2011, biochemists Jennifer Doudna and Emmanuelle Charpentier stumbled upon the properties of Cas9,[10] a protein found in bacteria that are already evolved to have the power to cut out viral infections. Doudna has described her discovery as 'fundamentally, a programmable protein that finds and cuts DNA'.[11] It could also be likened to the 'Ctrl-F' combination on a keyboard: a shortcut to finding the required typo in a very densely written book.

Using this protein it is possible for humans to artificially edit DNA sequences and modify the function of a specific, targeted gene for the first time, for instance in someone with a specific malfunctioning gene. Scientists take a copy of the gene in question, attach it to Cas9 and inject it into the person's cells. The coded protein will then seek out matching code in the body, and when it finds it, cuts it like a pair of scissors. The cell will attempt to repair itself but the process effectively disables the defective gene.

Just in case you were curious, Crispr itself stands for 'clusters of regularly interspaced short palindromic repeats'. This makes more sense when looking at a strand of DNA laid out flat: the sequence is populated by 'repeats' – the recurring A, C, G and T patterns – and 'spacers', which are bits of nonsense coding in between.

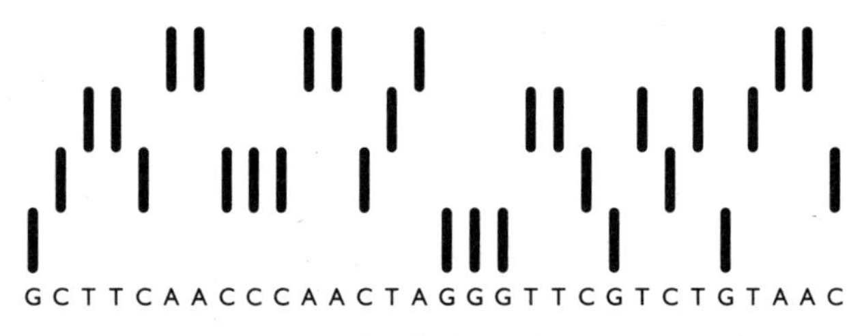

An example of a DNA sequence

Crispr has opened up new possibilities for scientists to 'correct' genetic defects and prevent the spread of diseases, and its application has the potential to span many other living organisms too. The tool could be used to edit out a genetic disease in livestock for example, or improve the growth of crops. Its potential impact is so widely recognised, that in 2020, Charpentier and Doudna won the Nobel Prize in Chemistry for their discovery.

Crispr has opened up the possibility of editing human DNA for health benefits and even aesthetic reasons – neither of which come without controversy. Unsurprisingly, it is at the root of a lot of the biggest questions society faces in our navigation of genomics, and will pop up as a topic of conversation throughout this book.

The promise in epigenetics

In mapping out the human genome, one of the big questions scientists had was, how does the information in our DNA translate to the instructions given to our body? As Stephan Beck describes it, the process is a little like running different computer programs using the same hard drive. 'The genome is the same in each of our cells but we run different genetic programs from the same genome in different cells,' he says. 'So, for example, your brain cells will run on a completely different set of genes than the genes your liver cells or your skin cells need to function.' In short, although we have the same genome in every cell of our body, each of our organs are affected by different parts of that genome.

So how do these parts of the genome send instructions to the relevant types of cell? The answer lies in epigenomes. We are thought to have around 100,000 epigenomes, which are chemical tags that give the specific and relevant instructions needed to activate or deactivate the genes in each specific type of cell. Another way of thinking about it is that the epigenetic marks that sit on our DNA are like traffic lights, Beck explains. 'So a brain cell would have lots of epigenetic stops, or red lights, on all the housekeeping instructions that control the metabolism, for example, because that's something the brain cells require less than liver cells. They would have the green lights on for genes that they need for brain function, however. And in the liver, you don't need any brain genes active, so the body silences them all epigenetically.' It's a very efficient way for the body to run its different genetic programmes from a single instruction manual, therefore.

The huge variety of epigenetic marks, or the unique combination of genes that are switched on and off, makes the number of their different combinatorial possibilities 'truly astronomical', Beck adds. 'We estimate it to be 10 to the power of 20 or bigger – and we don't even know all the epigenetic marks yet. New ones are discovered every year.'

The epigenome changes in accordance with our normal development from a single cell to fully grown adult – the tiny differences are all part of what makes each of us unique. Because epigenetic marks are very transient, they can be edited and removed using Crispr.

But the genome also changes naturally in response to our environment: there are specific epigenetic marks on the DNA of people who smoke or drink alcohol, for example, and these can be very telling of the impact of external factors. When Beck sequenced his own genome he found epigenetic marks indicative of tobacco smoking from when he smoked for a short time more than thirty years ago.

Because we have so many tiny and precise environmental and non-environmental changes marked on our DNA, epigenetics can provide a very useful biomarker for health and, more specifically, help to determine the cause of health problems in cases where the more obvious external factors leave little insight. The next generation of genomics will likely focus just as much on the epigenome as on the broader genetic picture, Beck believes. 'Epigenetics can be very useful in that we can screen large populations to ask, what is the effect of air pollution? What's the effect of smoking? What's the effect of diet on the epigenome? And if there are epigenetic changes, we can split those groups into people who have and have not developed a disease. And then you can ask the question, has this epigenetic change that is caused through lifestyle contributed to the disease as well?' he says. 'Epigenetics will help us to make bigger, better inroads into the prevention of disease and poor health.'

Applying genomics in the real world

Genomics is already solving a myriad of challenges faced by people and the planet for centuries. Take for instance the criminal justice system: the application of genome sequencing in forensics has opened up a door to solving countless fresh and cold cases that had previously baffled investigators relying on fingerprints and witness statements alone. 'It's almost impossible to commit a crime without leaving some DNA behind,' says Stephen Hsu, a geneticist and co-founder of a company called Genomic Prediction. 'Now we like to joke that it's just a matter of how much the detective wants to solve the case – because we can immediately find some family match.'

Hsu is one of several academics collaborating with Othram, the world's first private DNA laboratory set up to apply genetic sequencing techniques to DNA forensics. Based in Texas, the Othram team is a group of experts in the recovery and analysis of human DNA, using genomics to glean evidence and clues from the smallest traces of samples of degraded or contaminated materials.

Previous technology had been able to confirm whether DNA found at a crime scene matched an existing suspect's profile but now, Othram can generate a comprehensive genealogical profile from DNA remains found at the scene. This means that an investigator could know the likely age, height, ethnicity and sex of the offender even before they have any suspects in the case and so they could tailor their questioning accordingly. The same is true of unidentified bodies and victims: a genomic profile allows them to examine missing persons cases that might fit the bill.

The efficiency of the process is already coming as a shock to many aged criminals who probably thought they'd got away with it. 'We just had a confession from a 78-year-old serial killer who murdered a high school cheerleader after snatching her out of the car where she was with her boyfriend,' says Hsu. 'It's a case that's been unsolved since the 1970s, and now her family has justice, which is really incredible.'

Othram's work is just one of many examples of the potentially game-changing applications of sequencing technology. The fact that its results in forensics are so directly visible makes for an interesting case study, but most of the time the technology goes unnoticed in our day-to-day lives. Genomic sequencing is happening all around us, from crop modification determining the food we eat to the cars we drive becoming greener and the laws around us changing to match. And genomics is arguably having no bigger impact anywhere than in healthcare.

2

The health revolution

When she was a baby, Jess Spoor's parents knew that something wasn't quite right. She coughed and wheezed, suffered persistent chest infections, and struggled to put on weight. But regular asthma treatments, allergy tests and antibiotics didn't seem to help. Eventually, when Jess was two, doctors suggested they try a relatively new blood test that could screen for more debilitating conditions, just in case. The test revealed that Jess had cystic fibrosis (CF), a progressive genetic condition which causes a build-up of mucus in the lungs, breathing discomfort, digestive problems and, eventually, organ failure. Now thirty-three years old, Jess is one of around 10,600 people in the UK living with the condition. Her age is pertinent: the average life expectancy for CF patients in the UK is just thirty-one. In less economically developed countries, the outlook is much worse, and CF patients tend to die younger.

CF is caused by a defective protein that results from mutations in a gene called CFTR (cystic fibrosis transmembrane conductance regulator). One in twenty-five people are thought to carry the mutation, often without knowing it. When two parents happen to carry the gene, there is a one in four (25 per cent) chance that both of them will pass it on, resulting in a child born with CF. These days, parents of newborn babies are automatically given the option of screening for CF along with other genetic conditions through the commonly known heel prick test. But in 1989, when Jess was born, the disease was much less understood and the test was not yet routine.

Like most CF patients, Jess took tablets to help manage her symptoms throughout her childhood, periodically spending a few days in hospital when the mucus in her chest flared up. 'It was very much a part of my life for as long as I could remember, so I didn't really think too deeply about it,' she reflects. Thanks to

medication, a healthy lifestyle and regular chec
with her consultant, Jess's teens and early twenties were
fairly normal: she felt fit and well most of the time, went off to university at eighteen and later took up a job in hospitality.

'But things went downhill when I was twenty-seven and twenty-eight,' she says. 'I was suffering more and more and on antibiotics all the time. I couldn't work in an office environment anymore because the risk of infection was just too high.' Jess's medication was clearly no longer working for her; like many CF patients her age, she was running out of options. One summer morning, Jess was taken into hospital and assessed for a lung transplant, the first of several transplant assessments she would undertake over the next year. 'Which is when I think the reality of it hit me properly for the first time,' she says. 'I knew a transplant was something that would come up eventually, but it's really a last resort.'

Meanwhile, Jess's consultant was on to something. Daniel Peckham, who is also a professor of respiratory

University, had been monitoring the
drug that had been approved for use
at had achieved very exciting results.
ination of three drugs. Unlike previous
ich largely work to suppress symptoms
such as congestion, the drugs in Trikafta work together to target the defective CFTR protein and help it to function more effectively – meaning the patient's genes know what to do to clear the lungs naturally. 'We'd known about these drugs for some time,' says Peckham. 'The first in the trio was a drug called ivacaftor, and that was the first effective CFTR potentiator, meaning it could actually produce the missing protein.' Through studies putting ivacaftor together with two other, newer drugs, scientists had discovered a much higher efficacy rate. Trikafta was a winning combination.

Peckham likes to explain it using a farming analogy. 'Imagine you've got a field full of cows, and you need to get them in and out, but you can't do that very well because the gate is too rusty.' The cows in this scenario are the messages waiting to be sent to the brain; the rusty gate is the malfunctioning gene preventing the messages from getting through. 'What the drug does is open that gate – but the gate still has to be in the right place to start with.'

One of the challenges of treating CF is that not all gene mutations are the same – in fact there are approximately 2,000 mutations of the CFTR. One of these mutations, known as F508del, is particularly common. Trikafta had been shown to work in patients with at least one F508del mutation in their CFTR gene, who make up around 90 per cent of the CF population. Jess, Peckham knew, fell into this majority – and if she could get hold of it, Trikafta might just save her life.

When Jess went into hospital at the end of 2019, Trikafta was undergoing trials in the UK under the brand name Kaftrio, and wasn't yet widely available – but Peckham managed to get her a prescription on compassionate grounds. On New Year's Day 2020, Jess started treatment via IV in hospital, and the difference was almost instant. 'Within days I felt well again,' she says.

Jess says she never used to think about the future, and it wasn't something her family liked to discuss – the concept was just too uncertain. These days she has two jobs, one with her dad's business and another for the charity Cystic Fibrosis Care. For the first time, she is also, hesitantly, making plans. 'I'd like to travel more and keep working hard and doing well at what I do,' she says.

It's thanks to genomic sequencing that scientists have been able to not only detect which specific gene out of 20,000 is responsible for causing CF but also develop a life-changing treatment for sufferers – the oil for the farmer's gate. Patients like Jess provide clear evidence that single-gene mutations like CF will become increasingly manageable over time.

'Kaftrio is absolutely changing patients' lives,' says Peckham. 'We do need to be cautious, and there are still lots of people who are very sick. But already our hospital admissions have shot down. People are much more stable.' Peckham anticipates that CF will follow a similar path to HIV, in the sense that the disease – which was once a death sentence – could soon be one with which patients can live to a normal life expectancy with the right care.

There have been some unexpected outcomes of the treatment's rollout, too. In August 2020, Kaftrio was made more widely available in the UK through the National Health Service, and in the months that followed, multiple CF patients across the country discovered that they were pregnant. It's very rare for anyone suffering from symptoms of CF to be able to have children for a number of reasons, but one of the lesser-known complications of the disease is that it can cause infertility. Somehow, Kaftrio appears to lift this blockage.

Laura Gayton found out she was pregnant with her son Louis in November 2020, just seven weeks after starting Kaftrio. 'It was amazing, and a complete shock. As a CF patient it's not something you think is ever going to happen for you,' she says. Laura's husband, Nick, does not carry the CF gene, which means the disease hasn't developed in their child. Louis is a carrier, but the rapid progress taking place in CF research means that even if

Laura's descendants were to develop CF, they could live a long and healthy life.

More incredible still, a study of one Kaftrio baby found no evidence of the mutation being passed down. What this means, exactly, is something experts are still figuring out – but the turnaround from CF's devastating prognosis thirty years ago is considerable. 'I would expect Kaftrio to be extending life for this generation by twenty years-plus,' says Peckham. 'For children being diagnosed now, I believe we could expect them to live a normal life.'

Inheritance

When you consider how many letters of DNA each of us contains, it's hardly surprising our bodies don't always perform as they are designed to. Imagine, for example, if you were to type out all 3.2 billion letters of your genetic code – you would no doubt introduce a number of typos.

In fact, each of us has around four million little spelling mistakes in our genetic code. This is quite normal, and often harmless: the variation is what makes us unique, with different hair, eye colour, height and so on.

But every now and then, one of these typos can result in a change to our genes that has a significant consequence to the way our bodies form and function, and these changes can be passed on through generations. Examples of genetic conditions can range from the more structural changes such as cleft lip, which can be repaired through surgery, to more serious and life-altering conditions like CF, muscular dystrophy and some cancers.

As a professor of molecular ophthalmology and consultant ophthalmologist at London's Moorfields Eye Hospital, an important part of Mariya Moosajee's job is talking to patients and their families about inherited eye conditions and the impact of 'faulty' genes. 'One of the first questions they often have is, why did this happen?' she says. Whenever genetics is mentioned in connection with a health problem or degenerative condition, there is often misplaced guilt, 'the feeling of "Oh my goodness, did I do this to my child? Did I pass it on?"'

It's true that genetic health conditions can run in families: if one or both of your parents has a particular alteration in a gene, it can increase your chances of being born with it, or of developing the condition associated with it. Children won't always inherit the condition, however, and not all genetic conditions are passed down from parents – some gene changes can occur randomly while the baby is still a ball of cells developing in the womb.

Around 99.8 per cent of our genome is identical in everyone, leaving just 0.2 per cent that is different. But that 0.2 per cent difference equates to about four million letters. Of those four million letters, about 10,000 are potentially disease causing. Without genomic insight, whether or not you pass something bad on to your descendants is really a game of chance. 'So it's not something that you have done, or your ancestors have done, it's just the fact that, unfortunately, you are carrying this change,' says Moosajee.

Researchers believe this might not always be the case. Gene editing technologies are already opening up the possibility that one day a genetic disease such as CF could be eradicated. In 2021, a study published by scientists at the Hubrecht Institute in the Netherlands demonstrated that the defective piece of code found in CF patients could be swapped for a healthy piece of DNA instead, allowing cells to function completely normally. The researchers used a technique called prime editing, a similar alternative to Crispr, which allows experts to insert the new genetic material without causing damage to other places in the cell. The technique is yet to be tested out in humans, but initial studies proved successful in mice, and tests on mini-organs grown from the stem cells of human patients have also proven successful. The healthy DNA wouldn't be passed down to patients' children unless the edits are made to sperm and eggs, which is not currently allowed by law. But the findings are huge and provide the best indicator yet that cystic fibrosis could one day become an easily curable condition from birth.

Significant progress has been made in other genetic disease research, too. Sickle cell anaemia is an inheritable blood disorder affecting around 300,000 babies born each year.[1] Like CF, sickle cell is deadly and caused by a single cell mutation. Fortunately, recent studies using Crispr-like tools have enabled researchers to eliminate the disorder in mice. By changing a single letter of DNA in the red blood cells, it was possible to convert the disease-causing genes into harmless variants that make healthy copies of themselves. 'Our hope and expectation is that this approach might result in a durable, one-time treatment and potentially a cure for sickle cell disease that carries fewer risks to the patient,' said David Liu, a collaborator on the study.[2]

A similar story is true of some inherited eye conditions, and experts are hopeful about studies underway in research on HIV, Huntington's disease and muscular dystrophy – all of which are prime candidates for Crispr since they result from mutations in a single, identifiable gene. Eradicating genetic diseases from birth, if not before, is no longer just the stuff of science fiction: it could soon become a reality, if we allow that to happen.

Opening up genomics to all

It's likely that genetic screening will soon be commonplace in public healthcare – a routine part of signing up to your GP surgery for example might be to give a blood sample for your genome to be screened and added to the database. Experts are divided on when the most appropriate moment in a patient's life might be to get their genome screened – it could be when they reach the age of majority, or when they first present with a medical problem. Others would push for a full genome screening to take place from birth, or even earlier.

At Moorfields Eye Hospital, Moosajee's patients are already benefiting from sequencing technologies. Most of the people who are referred to her have some kind of serious condition that is probably genetic, and she offers them all the option of having their whole genome sequenced. A simple blood sample is taken, or saliva if the patient is particularly needle-phobic, and sent to an accredited NHS genomic laboratory based at Great Ormond Street Hospital nearby. 'We can extract the DNA from the blood cells and use sequencing to understand and identify which particular mutation a patient has – which in turn will confirm the best course of management,' says Moosajee. The whole process is quick and effective, and the patient's DNA can be stored indefinitely with their consent, which can be useful for consultants should they need to perform further tests at a later date.

However, the process doesn't come without its emotional challenges. 'There has to be a discussion first about the ramifications of patients finding out that something is a genetic condition,' she says. Patients need to be prepared for the possibility they could pass on a faulty gene to their children, or that they already have done, for example. Sometimes, the opposite problem occurs: genetic testing can reveal mistaken paternity or other family secrets which patients are not always prepared for. But genetics counsellors are on hand to support patients where needed.

Sequencing on the whole has made 'a fantastic difference to the way we work', says Moosajee. 'Traditionally, the tests we did enabled us to look more generally at big panels of genes, which equated to an average diagnostic rate of about 25 per cent for all genetic eye disease patients put together. That meant 75 per cent were not getting a diagnosis even though their condition was genetic. Now we have whole-genome sequencing, accurate diagnoses have more than doubled, taking the average up to 56 per cent.'

Once too slow and expensive for general healthcare, whole-genome sequencing is becoming normalised in hospitals and treatment centres. As such, it is opening up pathways to precision treatments and better care because it allows doctors to find changes that were never previously looked at or were outside of the relevant gene. 'Before, we were only able to pick the low-hanging fruit, but now we're looking at every single letter to find mutations,' says Moosajee.

In May 2021, the UK government announced it would make whole-genome sequencing available to patients through the NHS as part of routine care services, making it the first national healthcare provider to do so.[3] The NHS Genomic Medicine Service is already available for patients with rare diseases, including genetic eye disease, suspected cancer and other serious conditions. It's hoped that the service will help to reduce waiting times and adverse drug reactions.

That genetic sequencing is now so widely available to the public is thanks to work started by the 100,000 Genomes Project in 2012. Launched by the then prime minister, David Cameron, whose own son Ivan died from a rare genetic disorder, the programme set out to sequence whole genomes from NHS patients for the benefit of genetic research. The work was declared complete in 2018,[4] but it set a precedent for whole-genome sequencing for wider public healthcare. Stephan

Beck set up the Personal Genome Project (PGP-UK) in response to the 100,000 Genomes Project, which was limited to sequencing the DNA of very sick patients who met strict criteria. 'We wanted to open up the service to everyone, so we said everyone can do it [have their DNA sequenced], but in return they have to donate their data so that it's openly available,' he explains.

PGP-UK is based at University College London and run by a wider team of researchers. But it is part of an international network of genome-sequencing projects (the Global Network of Personal Genome Projects) set up by Harvard Medical School researchers with the aim of creating freely available scientific resources from genomic data. Open sharing of genetic data is one of the founding principles of the project, the idea being that the more people can benefit from the resource, the more the research can evolve to provide quality analyses of our disease risk and traits. The genomes of the public are anonymised but, even so, Beck says he and his colleagues were pleasantly surprised by the enthusiasm received for PGP-UK. 'We announced the launch of it on the radio one morning, and by the end of the week we had more than 10,000 volunteers. There's personal gain for them in doing it [getting the results of their personal genome sequence back] but there's also a big altruistic element that people just want to do something for the better of humankind,' he says.

There's another benefit to the cause, too: Beck hopes it will open up genomics to a much more diverse dataset that is truly representative of the whole population. 'We have a very big issue in that currently the majority of the data that make genomic medicines work, and that our calculations are based on, only apply to Caucasians,' he says. It's a scientific fact that there are many genetic differences between ethnic populations, Beck explains. 'And each of these differences contribute in a different way to things like disease susceptibility, how an illness presents itself, and sometimes in how well the body reacts to treatments, too.' So far, the vast majority of people who have had their genomes sequenced are white and European. That means the vast majority of reference data in public databases used by researchers working on new therapies continues to be representative of a narrow cohort of people with European ancestries, creating a divide between those who benefit from genetic sequencing (primarily white people) and those who don't (people of other ethnicities).[5]

The reasons for this data divide are many, Beck believes, but the solution requires better communication and public education of the value in genomics, as well as recognition of this data bias by the funding bodies that finance sequencing projects. As evidenced by vaccine take-up and organ donation figures, certain communities are more likely to lack trust in the state and have no personal incentive to give up their data. 'It's something we have to work on,' Beck says, 'because if we want to offer genomic-based treatments and diagnoses to the wider population, it has to be fair. And fair means that the data that are currently missing would need to be generated.'

Where genomics – and gene editing – has its limitations

Tools like Crispr offer what seems like a relatively straightforward solution to so many genetic diseases, which poses the obvious question: why is the technology not being used to cure all patients already? The reality is that the relative newness of Crispr makes a lot of people wary about the possible long-term effect of its use in humans. It is, in any case, not without its limitations, at least at present.

A major risk is that Crispr's 'scissors' have sometimes cut DNA outside of the intended gene. This is known as 'off-target' editing. Such unintended edits are thought to be rare, but could have significant consequences, such as turning cells cancerous.[6] Consider that the human genome is more than three billion base pairs of DNA in size: currently, researchers can put in an 'address' for the Crispr Cas9 protein to find, but the protein itself is the equivalent of twenty nucleotides – or building blocks – in size. There is a possibility, therefore, that once it reaches its destination, the message could overlap into different areas in the genome, causing unintended cuts and reorganisation of the cells. It's a concern for Maarten Geurts, co-author of the Hubrecht Institute study – and something he's seen first hand in the lab. 'Off-target editing is very dangerous – genes can start to duplicate and we don't necessarily know how to fix that,' he says.

There are historic examples of genetic manipulation gone wrong, too, which offer a stark warning of the dangers involved. In 2002, a trial led by French researchers at first appeared hugely successful in demonstrating the use of gene therapy to cure patients of severe combined immunodeficiency disorder (SCID), an inherited immune deficiency also known as 'bubble boy' disease. But the trial was suspended after two of the patients involved developed symptoms of leukaemia – one of whom died.

Gene therapy is slightly different to gene editing (in which a mutated gene is revised, removed or replaced), as it attempts to offset a mutation by the insertion of a healthy version of the same gene – the disease-related genes will still remain in the genome. The bubble boy study took place before Crispr tools came about, and scientists' understanding of gene therapies has improved since that time, but the study's legacy means that many in the field remain uneasy about utilising gene editing in patients.

Prime editing, which was invented by David Liu (who was a collaborator on the sickle cell study), is used by Geurts's team in the CF study and offers what many believe to be a safer alternative to Crispr. Geurts describes it as 'Crispr 2.0: we use a version of Crispr which is able to find the target gene without cutting it completely,' he explains. Where the traditional Crispr tool cuts both strands of DNA, prime editing uses a modified version of the Cas9 protein to make only a single strand cut, reducing the risk of unintended damage.

The method is one of many new variants of gene editing technology which hold great promise in the journey towards making such therapies safe for common use. Geurts is often contacted by members of the CF community anxious to hear more, but he cautions that the process won't be available for use in human patients for a while – the technology will take time to be approved for human use. 'It's difficult but important that we give patients a realistic view of where we stand – in vivo trials are being done [trials using living organisms and not just cells in a dish] but we have to work on safe delivery of these proteins,' he says. 'It doesn't mean that within the next two years we're going to have a cure. But within the next five years we will start to see clinical trials of new ways of targeting cells in the lungs, delivered through inhalers for instance. That I am 100 per cent sure of.' In the meantime, pioneering new treatments are being made available thanks to the knowledge gained from sequencing studies about how the disease works.

Outwitting cancer

Perhaps one of the biggest areas to be impacted by genomic sequencing in our lifetime is cancer research.

Cancer is a genetic disease that occurs when mutations in DNA cause cells to divide and grow uncontrollably. Scientists have identified some mutations already – a well-known example is BRCA1, a gene causing breast cancer, and some treatments already exist to target specific gene mutations like it. But current predictions suggest more than 200 different types of cancer exist and, given the unpredictable and often evasive nature of the disease, finding the best way of treating each one can seem a bit like looking for a needle in a haystack.

One way that researchers hope to mitigate this is by mapping out every type of cancer cell to gather data on all cancer's weaknesses. The Cancer Dependency Map (DepMap) is an international project led by the Wellcome Sanger Institute alongside the Broad Institute at MIT and Harvard. The first element of DepMap is to create new cancer models – clusters of cells that recreate patient tumours but can be grown in a dish and studied in the laboratory. More than 2,000 such models have been created and uploaded to the database so far, but researchers hope to fill in the gaps using Crispr and similar technologies. Beyond helping scientists to understand cancer's weakness, these models also allow experts to safely test and identify the effects of different drugs on each particular cell type without having to experiment on patients themselves.

Because the consequences of alterations in the DNA of cancer cells are not fully understood, project leader Mathew Garnett hopes to assign a 'dependency' level to every recorded cancer – that is, a marker of all the mutations which make each cell vulnerable to becoming cancerous. The hope is this will better equip researchers to develop precision treatments going forward. 'There is no question in my mind that genomics is dramatically changing the way we treat cancer patients today,' says Garnett. 'We can use genomics for prognostic reasons, for example if we identify a genetic change through sequencing we can tell if a patient is more or less likely to respond to a treatment compared to someone who has a different set of genetic changes.' In other words, a patient's genetic make-up, as mapped out by sequencing, can provide a reasonable insight into how their body might react to different treatments.

Sequencing also allows researchers to see which genes need to be targeted in designing future medicines. This is where Crispr plays a role: using the cancer map database, researchers have used the gene editing tool to disrupt more than 18,000 individual genes in the genomes of 324 different cancer cell types (taken from patients and grown in the laboratory). These 324 cells represent thirty different types of cancer. They are stored in the lab and grown into a population of the same type of cells, known as cell lines, creating a living database for researchers. 'If a cell line stops growing, we can classify the gene that had been disrupted as essential to that cancer's survival,' Garnett explains. In other words, through a process of elimination researchers can identify those genes that support the cancer's growth and, therefore, how we can better stop cancers in their tracks.

Since starting the project, Garnett's team have discovered thousands of 'essential' genes across all the different cancer types – which opens up the possibility of finding safe ways to remove them through targeted treatments. The final stage of DepMap will be to match the cancer models grown in labs with datasets from real patients, which will link genetic changes in particular patient groups to the drugs that work.

All this means that one day before too long, cancer patients could receive personalised treatment plans based on their own genes and those of the cancer they have as a matter of course. Helping doctors get to that point is a field of research called pharmacogenetics.

Pharmacogenetics explained

Researchers in California are sequencing genomes in order to draw up a database of how bodily variations, like metabolism, can impact the efficacy of drugs. The idea is that eventually patients will be able to get their genome tested and compared against the database to determine their prescription: which particular cancer drug might work best for them, for example, and what dose. This would reduce unnecessary toxicity, creating a more pleasant treatment experience for them, but also save time and resources, reducing the likelihood of wasted efforts on the wrong drug.[7]

As Stanford University professor Russ Altman describes it, pharmacogenetics is really just a shortcut towards what doctors already try to learn through many years of experience: what might work best for a certain body type, shape, ethnicity or medical history. 'Drugs are approved, but they are approved for the average person, and they're approved to be generally safe,' he explains. 'But there is huge individual variability, as anybody who's ever gotten a prescription will know. Doctors will prescribe a drug and say "This drug is going to do *x*", and sometimes it does *x*. But sometimes it doesn't, and sometimes it does *x* plus a lot of other things that you weren't expecting.'

There are many environmental reasons for variability in drug response: what your diet is like and what you've been eating that week, what chemicals you may have been exposed to. But another main determinant is your genetic background, which means your response to a drug is largely inherited from your parents and grandparents. This knowledge should really simplify things, but while GPs will often ask their patients for any family history of diseases like Type 1 diabetes or glaucoma, 'there's very often no family folklore about how Grandma took this drug and didn't do well', says Altman.

'The vision', he adds, 'is that we measure the genome and then we use that information to advise physicians about which drug is most likely to work for a patient, almost in the background. So neither the physician nor the patient should have to think about it explicitly, because the clinical information systems are just considering all of this and making their recommendations.'

With enough data, the work could be reduced to an algorithm. So far researchers at Altman's lab have added around 1,000 drugs with some known genetic influence to the public pharmacogenetic database – no small feat. Altman prioritises the drugs that clinicians tend to understand the least, meaning the ones which vary the most in their reception. 'Some cancer drugs are miraculously effective for some people and really very ineffective for others,' he gives as an example. 'That's a canary in a coal mine for potentially a genetic influence on the response.'

The limits of genomics in cancer research

Ultimately, even with all the mapping and genetic insight in the world, cancer is such a complex and inherent part of our natural evolution that some believe it will be impossible to eradicate the disease entirely. 'While we have had some huge successes in the field, unfortunately limitations are brought about to some extent by the ability of cancer to evolve and evade therapies through acquired resistance,' explains senior oncologist Charlie Swanton. 'Even if we knew everything about the cancer genome and risk, even if we knew everything about the patient's germline [the cells which pass genetic information down] it can't necessarily tell you that a patient will or won't get cancer, because chance still plays a role. You can develop cancer just through pure bad luck because of the way in which a cell divides or the mutation it acquires due to ageing. It's very hard to anticipate how we can overcome that problem.'

Cancer is, for now at least, too vast and varied a disease to target with a precision tool like Crispr. A tumour can be made up of as many as 10 billion different tumour cells and can be spread across multiple sites in the body. An added complication is that, just as species like humans and animals evolve to adapt to our environment for the best chance of survival, cancers also evolve to become resilient to treatments. As such, Swanton explains, 'right now, our genomics tools are still not sensitive enough, so we can't know what the extent of variation and diversity is in a tumour that will enable you to predict how drug resistance occurs. We're a long way from understanding what's going on in the genome at the level of protein production, and at the level of the epigenome, how the genome is regulated in every cell, in every tumour, in every patient. It's probably something we'll see in our lifetime. But right now, even if we could see, read and understand every tumour cell, we'd have a problem accessing every cell at every site of a living person's body.'

Opening doors to new kinds of treatment

This dilemma has led Swanton and his colleagues to think outside the box. Through the PEACE study at the Francis Crick Institute in London, researchers are sequencing tissues taken from the dead bodies of patients who volunteered themselves when alive and were faced with terminal cancer. It's hoped that by having access to the whole body, researchers can further their understanding of how and why cancer spreads the way that it does, and why and how the disease's trajectory varies so much between patients. Their findings have already shown indications of genetics playing an important role. 'We're trying to use genomics to improve outcomes. The difficulty is, of course, that for the patients taking part in PEACE it's all too late,' says Swanton. 'But we're hopeful that PEACE will help unravel some of those secrets for the care of future patients, certainly.'

Cancer patients need a better quality of life while they are living with the disease, too, and researchers are prioritising new leads in therapy development. It's hoped that even without an ultimate cure, precision treatments could extend and improve the lives of patients diagnosed with even the most vigorous of cancers.

One new kind of treatment is already offering huge promise to previously untreatable cancers. Immunotherapy, also known as immuno-oncology, is a form of cancer treatment that uses the body's own immune system to destroy cancer cells. It comes in a variety of forms and can be used either on its own or in combination with surgery, chemotherapy, radiation or other targeted therapies to improve their effectiveness.

Immunotherapy has existed in various forms for around a century, but genomics has given the field a turbo-boost. In 1891, a doctor called William Coley began experimenting with injecting bacteria into cancer patients to encourage the body to fight off the infection, and with it the cancer. His methods had limited proven effect, but the idea was not dissimilar to the theories put into practice now.[8] These days, scientists use drugs called checkpoint inhibitors, which are genetically engineered cells that effectively release the brakes on the body's immune system, spurring it on to fight back against dangerous mutations – for instance, when a cell starts to multiply to form a tumour. It doesn't work for every patient, and some immunotherapies can be associated with unpleasant side-effects, just like traditional drug treatments for cancer. But scientists are excited about its potential as a gateway to more personalised, precision treatments.

'The crazy thing is, we don't fully know how immunotherapy works,' says Swanton. 'We probably understand maybe a half to two thirds of the way in which the cells act, but there is still a missing component. We don't know where that component is. It could be in the genome, it could be in the germline of the patient, it could be in the cancer genome, or it could be coming from elsewhere. We haven't got a full picture yet. Genomics will be fundamental to helping unravel that, without a doubt.'

One immunotherapy treatment to have emerged from Swanton's lab at the Francis Crick Institute has been developed into the primary treatment at Achilles Therapeutics. Achilles is a biopharmaceutical company developing personalised cancer therapies that work by targeting protein markers found in tumours called clonal neoantigens. Clonal neoantigens are only present on tumour cells and are not found in healthy tissues, enabling scientists to specifically target them for therapy while leaving healthy tissue unaffected. In this way scientists are exploiting the cancer's 'Achilles heel', using its own tricks against it.

To conduct the therapy, DNA from the patient's tumour cells is extracted after the tumour has been removed during surgery. DNA from the tumour is then sequenced to provide an exact genetic signature of the cancer. Healthy DNA taken from patient blood samples is also extracted and sequenced for comparison with the tumour's genetic code. From this, experts can identify mutations that are present on the tumour but not on the healthy tissue. 'That leads to the identification of a specific set of proteins that we can target with the immune system,' explains Iraj Ali, Achilles's CEO. The second step is to extract T-cells from the patient – that is, white blood cells or immune cells, extremely potent killers of unhealthy cells – in order to enable researchers to effectively supercharge them for use against the cancer.

Research into the natural evolution of tumours, helped by genomic sequencing of tumour tissues at the Crick, has already highlighted one of the underlying problems in cells that make tumours grow and spread across the body. Genomic studies have led researchers to believe that immune cells effectively get 'stuck' inside the tumour tissue and become dysfunctional and shut down. But by extracting those dysfunctional immune cells, and reactivating them in the laboratory, researchers can return highly functional and more specific immune cells, selected for their abilities, into the patient's bloodstream. 'We are reconstituting the patient's own anti-cancer immune system to produce an individualised, precision therapy,' says Ali.

Cancer can be thought of a bit like a tree. Its trunk – the core DNA mutation – is formed by clonal neoantigens, those proteins I mentioned earlier. Later mutations will only be present in a subset of cancer cells, which are known as sublocal or branch mutations – the branches of the spreading tree. The problem with traditional cancer drugs and treatment is that they tend to target the branches and leaves, by which point the cancer has evolved and potentially gained some resistance to the drugs. Using the immune system as the body's own defence targets the trunk of the tree, which helps to prevent resistance from occurring in the first place.

This form of immunotherapy that targets clonal neoantigens is still undergoing trials, but researchers are encouraged by the results seen in patients so far. Funding such an initiative requires millions of pounds, Swanton explains, 'beyond anything a funding agency could support', which is why Achilles has been established as a public limited company with backing from investors. Phase two trials are underway in patients with advanced lung cancer and patients with metastatic or recurrent melanoma, and it's hoped that the treatment could be rolled out to patients across public and private healthcare within five years, subject to regulation and pricing approvals.

The preliminary success of immunotherapy trials like these not only gives promise to patients for whom cancer drugs have proven harsh and unrewarding, it also marks a shift in the way scientists are thinking about cancer itself. 'The Holy Grail is now trying to identify therapies that are really going to cure cancers that have spread beyond the primary site rather than just keeping them at bay,' says Swanton. 'There's been a missing link there for a long time, but we're hoping that immunotherapy will provide the key,' he adds.

Paving the way for better healthcare

For decades we have relied on medicines such as antibiotics to help us lead healthier and longer lives. And antibiotics have served us well: if you get an infection of almost any kind, there's a prescription available to clear it up, often in a matter of days. But there is a huge problem looming over us ready to throw that up in the air.

Antimicrobial resistance (AMR) is when infection-causing bacteria become hardened to the drugs that are meant to kill them. It's a process that has been taking place gradually since antibiotics were introduced, and our overreliance on them is partly to blame. Speaking on the issue in 2017, the UK's former chief medical officer Sally Davies warned of an 'access versus excess issue' in that some rich countries have easy access to antibiotics and have a tendency to overprescribe them. In England, as many as a fifth of prescriptions for antibiotics are unnecessary, according to public health medical directors.

The problem boils down to the fact that bacteria are evolving more quickly than our development of new drugs against them: more than 700,000 people globally die each year from infections caused by antibiotic-resistant bacteria. Meanwhile, all new antibiotics brought to the market since 1984 have been variations of the same drug. If the problem continues, we will no longer be protected against serious infections like blood poisoning and pneumonia. Operations and births could once again become the life-threatening endeavours they were in medieval times.

Crispr and similar gene editing technologies offer a new line of research which could help to solve the problem by opening up entirely new kinds of treatment and protection against bacteria. At the University of Oxford's Department of Zoology, scientists have learned through genomic sequencing that bacteria use multiple mechanisms to evolve to dodge the effects of antibiotics. But now, researchers are gaining ground on bacteria once more. 'It's possible we could experiment with editing the bacteria genome to help us develop more effective anti-infection drugs than we have now,' says Celia Souque, an Oxford microbiologist who uses lab experiments alongside genomics and mathematical modelling to study AMR. 'We definitely need new antibiotics, but genomic technologies are also creating a lot of space to try new approaches, like stopping bacteria themselves from evolving,' she says.

***

AMR preventative technologies, gene editing, advancements in immunotherapy and the genetic sequencing of cancers are just some of the clear examples of how genomics could revolutionise everyday healthcare. Genomics could solve some of the most complex challenges in medical research as well as some of the most high-profile and dangerous diseases that humans face. It could radically change the way we think about healthcare, open up the possibility of precision medicine, help us to lead healthier lives – and even help us to combat the threats we don't even know exist yet.

3

Protecting against future threats

Viruses are by their very nature unpredictable. Classed as neither living nor inanimate, these infectious agents survive by their stellar ability to mutate quickly, dodging their primary predators – immune cells – with rapid shape-shifting techniques to keep them disguised and help them spread. For this reason, viruses are notoriously difficult to contain or control, as everyone learned following the outbreak of Covid-19.

In autumn 2020, after a summer of socially distanced drinking, dining and general making up for lost time by the UK public, health experts were growing increasingly concerned by the number of Covid-19 infections in south-east England. A second national lockdown had been put in place, restricting people in England from leaving home unless truly necessary, and yet cases were still rising. The problem was especially noticeable on the tiny Isle of Sheppey, 93 square kilometres in area, off the mainland in the borough of Swale, Kent. During the first wave of Covid, Swale's infection rate had been relatively low, but this time the numbers were shocking: by 18 November, Swale was the UK's virus hotspot with 631.7 cases per 100,000 people. On the east of Sheppey, where three prisons are located, the rate was at 2,079.5 cases per 100,000 – around nine times the national average.

With no vaccines yet available, government ministers were left scratching their heads as to why their rules weren't working, but also at what to do next. Speculation grew, particularly when it came to trying to work out what was going on in Sheppey. Perhaps people there weren't taking the restrictions seriously? Were the prisons themselves to blame?

Calling an emergency meeting on 23 November to address the rising case numbers, the leader of Swale council, Roger Truelove, expressed his frustration at a 'wilful disregard of the rules'. He would 'write to supermarkets' to warn them against overcrowding, and 'supercharge the messaging' about social distancing rules to the general public, he pledged.[1] He wasn't to know that the people of Kent had been dealt a particularly unlucky hand.

SARS-CoV-2 – the virus causing Covid-19 – is relatively slow to mutate. This is good news for us in some ways, as it makes it easier for researchers to design drugs and vaccines against it. But it also makes the virus more difficult to track, because so much of each new variant's coding is identical. With much of the genetic make-up of the virus unchanged, it can be difficult to spot the small new mutations taking place. However, by the time of the outbreak in Swale, experts knew that the genetic make-up of SARS-CoV-2 had probably changed; that was to be expected, given the natural evolution of viruses. What the data suggested to them was not that people in Swale had developed a more careless attitude, but that something significant had now happened to the virus's genetic code that made it even more transmissible than before.

When sequencing the genome from patient samples in Kent, researchers spotted an unexpectedly large number of genetic changes in the virus's code. In one big evolutionary jump, the new variant B117 had picked up twenty-three mutations, eight of which were on the spike protein, which helps the virus lock on to human cells. This was significant because it potentially affected how transmissible the virus was. Humans were going to need to up their game, which in the short term meant two things: first, that Christmas holiday plans for much of the nation were cancelled that year, and second, that increased genomic surveillance was required for experts to keep on top of the virus's every move.

'People say, well, couldn't you have stopped B117? Actually, I haven't ever seen anybody anywhere be able to stop a variant travelling once it emerges,' says Sharon Peacock, executive director of the Covid-19 Genomics UK Consortium (COG-UK). 'You can't put the genie back in the bottle. But you can learn how to prepare for the next one better.'

Like many people in her field, Peacock first learned of the novel coronavirus later identified as SARS-CoV-2 very early on through conversations with international peers around the turn of the 2020 new year. She knew that there was urgent work to do. Peacock, a leading microbiologist at Cambridge University, put out a call to five trusted colleagues in the sequencing community to say, 'Can you give me a ring back? I think we're going to need to set something up quickly.' In March, Peacock held a round-table workshop of experts scattered across the UK who, before the pandemic, worked with genetic sequencing in other areas. 'There was no fancy scientific presentation, we just said, "How on earth are we going to do this?"'

'We had certainly been working on genomics in public health microbiology in the decade leading up to Covid-19, including its use to investigate other outbreaks,' Peacock says. 'That set us in good stead for Covid-19 in that we had the capabilities, knowledge and expertise in place ready to call on.' It would be a stretch to say that experts were prepared for an outbreak of Covid's scale, she admits, but they were certainly ready to set aside what they were doing and prioritise the new global threat that stood before them. The group quickly built a network of sequencing labs, taking into account the volume of data on Covid that would need to be collected, but also geography – there was no point in sending samples far and wide to get sequenced, especially when time was of the essence and travel was so restricted.

By 1 April 2020, COG-UK was up and running, ready to deliver large-scale and rapid whole-genome sequencing of the new virus to monitor the impact, scale and speed of its spread. Genomic data from the patient and virus samples, together with epidemiological data – people's movements, for example – provide the best possible clues to scientists in terms of the routes of the virus's transmission. Gathering data from patient samples across the country, COG-UK researchers were able to help predict the path of new variants, which fed into government guidance such as mask wearing and school closures.

It was COG-UK members who, on 18 December 2020, published details of the B117 variant, warning, 'The rapid growth of this lineage indicates the need for enhanced genomic and epidemiological surveillance worldwide and laboratory investigations of antigenicity and infectivity.'[2] By late December, the UK was responsible for roughly half of all the world's genome sequencing of the virus.[3]

After millions of deaths, not to mention the economic and social fall-out caused by the pandemic, it is easy to feel despair at the virus's impact. But without genomic sequencing technologies, there is no doubt we would be in a much, much worse position than we are now. Consider how, in the fourteenth century, bubonic plague caused an estimated 25 million deaths in four years in Europe alone. Without the sequencing tools at our disposal today, it's possible that we would have spent years in and out of lockdown.

'If SARS-CoV-2 emerged even fifty years ago, I don't think we'd be in a particularly informed place,' says Peacock. 'We wouldn't have been able to develop the vaccine so quickly for a variety of reasons, including less advanced vaccine development pipelines, but also because vaccine development has benefited greatly from the availability of sequencing technologies and viral genome sequence data, something that we have come to depend on in this century.' Having the genomic blueprint for any virus and its many variants allows vaccine developers to target that virus more specifically, because they can see its strengths and weaknesses as well as what it is in its genetic make-up that makes it so adaptable and transmissible. Part of the reason SARS-CoV-2 spread so rapidly was because it was unusually well adapted to live and spread in humans, but that knowledge enabled vaccinologists to design the best possible products to inoculate the human body against it.

SARS-CoV-2 may be the most surveilled virus in history. Crucial to this has been the willingness among researchers to share their data openly. On 5 January 2020, virologists Zhang Yongzhen and Edward Holmes published the entire genome of the virus from a patient admitted to hospital in Wuhan on 26 December 2019. Since then, at least 10 million SARS-CoV-2 genomes have been sequenced and uploaded to GISAID, an open platform for sharing viral genomes. More than 2.5 million of those genomes came from the UK, which sequences roughly 12 per cent of all its positive Covid-19 tests.[4] 'It's extraordinary to imagine how different our response would have been if we didn't have genomic data, which is being used to track the evolution of the virus [and] emergence of variants of concern, and redesign vaccines on an ongoing basis,' says Peacock.

Genomic surveillance

Tracking viruses was a full-time mission for a large team of people even before Covid-19, as the Wellcome Sanger Institute researcher Sonia Goncalves knows. Every day, her team at the institute's parasites and microbes facility takes in a delivery of samples ready to be sequenced from patients who have tested positive for malaria, with the aim of identifying mutations.

When she got the call about Covid-19 in early 2020, Goncalves knew things were about to get hectic. 'It was very intense,' she recalls. 'Although we had the sequencing capacity and the facilities in place, the issue with SARS-CoV-2 was that it required a completely different process, not only in terms of the lab methods, which had to be adapted and optimised for the new virus, but also in terms of how we were going to receive the samples.' Within a matter of weeks, a system was in place in collaboration with COG-UK for tens of thousands of samples from PCR tests to be sent from across the country to Goncalves' lab each week for testing. During times of high infection rates, the lab sequences around 4,000 genomes from the samples every day, each one adding to a live picture of where the virus is in its evolution. The process is entirely automated, Goncalves explains, which allows researchers to operate on a large scale: 'We developed analytical pipelines that automatically screen the data that comes out of the sequencing machines and detect changes when compared to a reference genome.'

Goncalves never imagined that surveillance of SARS-CoV-2 would be a priority for this long – she started in 2020 with a six-month plan, but that has since been extended to cover at least two years. What comes next is a long-term surveillance programme to monitor the mutations of SARS-CoV-2 long into the future. 'Even if the pandemic reaches a controlled state, we will need to have kind of a baseline genomic surveillance just to understand how the virus is still evolving – which it will,' says Goncalves.

Rather than raise alarm (yes, experts do believe it's likely we will be living alongside this coronavirus forever), the long-term planning should provide reassurance that experts stand a better chance of spotting new waves of infection as the virus continues to mutate. It will also help to inform future policies around winter planning. Long-term surveillance of flu virus variants, for example, enables experts to predict winter outbreaks and make a judgement call on which flu vaccines to roll out each year.

Before Covid-19, Goncalves worked on malaria sequencing full time for similar reasons. 'We run a network of programmes across different countries and regularly sequence sets of samples that they send to us, then we provide information back to each programme,' she explains. 'It helps us to understand how the parasites are evolving in terms of drug resistance, and also the effects of measures used to control the mosquitoes which carry the virus.' The need for surveillance is constant because new mutations are always popping up, she adds. 'It gives us the ability to detect them as soon as they appear, so we can immediately communicate a change to the relevant agencies, who can then use that information to support their decisions on how they're going to control the outbreak.' If new malaria sequences flag a high frequency of resistance to a specific drug in a specific region, for example, the control programme in that region may decide to switch incoming patients to a different drug to give them the best possible chance of forming antibodies against the virus. 'These are the kind of decisions we take on a daily basis,' says Goncalves.

In the future, she predicts, genomic technologies for virus surveillance could become cheap and small enough to be used like any basic app on a mobile phone. 'You could go into a room, for example, and using genomics technology, your phone will detect a virus is present in the air. We are very far away from that, but you can see how people are already thinking about the next step, and it's very exciting.'

4

Navigating the challenges of climate change

It's not just killer viruses that are threatening our existence on this planet. Surely one of the greatest threats to humanity and the world as we know it is the devastation wrought by climate change. Unfortunately for us, rising sea temperatures cannot be countered with a single tonic, and the speed at which the planet is changing is undeniably, disastrously made worse by the actions of humankind. But genomics is providing positive progress here, too.

In agriculture, for example, genetic modification is providing solutions to farming challenges and resultant food shortages across the developing world. Increasing temperatures, droughts and flooding are making farming in Africa increasingly difficult, but since 1996 programmes across the continent have been breeding crops and livestock to have certain climate-adaptable traits, and with proven results.

To create genetically modified (GM) crops and animals, scientists will typically remove a desirable gene from one organism and introduce it into another. Grains like rice and maize, for example, are modified for growth in many sub-Saharan African countries to require less water and be tolerant to drought. Tuberous vegetables like cassava and potatoes are made resistant to common crop-failure-causing diseases.[1] The success of these new superplants and grains has helped to prevent famine and supported vulnerable economies in sub-Saharan Africa and other parts of the world. But there are environmental benefits, too: according to industry-based research, GM crops have reduced CO_2 emissions equivalent to taking 16.7 million cars off the road for a year because extra energy and water are not needed to maintain them in inhospitable environments.[2]

In the Western world, and in Europe in particular, GM farming has had less appeal – partly as a result of these nations' immense privilege in not suffering the same degree of crop failure as the Global South, but partly due to a general sense of distrust of seemingly 'unnatural' processes, not to mention a distrust of GM companies' motives. And there has been plenty of negative press surrounding GM to fuel these fears, particularly in the early days of GM farming. In 1998, Hungarian-born British biochemist Árpád Pusztai shared research allegedly showing how GM potatoes had stunted the growth of laboratory rats, and even damaged their internal organs. GM food has the potential to be harmful, he concluded, suggesting that much more research was needed to fully understand the link between genetic modification and toxicity. The findings understandably caused a huge stir in the media, but were eventually discredited.

So far, scientists have been unable to find any link between long-term health problems and GM food, and the official position endorsed by the World Health

Organization is that GM food is safe to consume. Calls for regulation of GM food grew with public awareness of its existence, and these days many countries require GM food to be labelled accordingly. But this hasn't put to bed all of the concerns. Environmental campaigners, among others, warn of the negative effect of genetic modification on biodiversity – GM farming is a kind of selective breeding, after all, and narrows the gene pools of some species. Even in 2015, more than twenty years after GM crops were first approved in the US, half of the European Union's twenty-eight countries voted against a scheme to allow GM crops as part of a wider initiative with a network of biotech companies, effectively making two thirds of Europe GM-free.[3]

In reality, humans have been altering the look, taste, size and content of plants for food and aesthetic purposes for centuries – and farming is the biggest example of that. The oldest evidence of artificially selected crops dates back as far as 7800 BC.[4] The only difference today is that genomics is offering a shortcut, and scientists can use gene editing tools to alter almost everything, from determining the colour of a flower to speeding up the ripening of tomatoes.

Attitudes are gradually shifting once more, especially as messaging around the impact of climate change on food supplies hits home. In 2009, the United Nations estimated that food production will need to increase by 70 per cent by 2050 if we are to feed the estimated two billion extra humans who will live on the planet by then.

On 29 September 2021, the UK government agreed a law change that would enable genetic modification in crops in England once more, but it also went a step further in allowing gene editing of crops. The move was met with some cynicism as it came soon after the country's departure from the EU and coincided with disruption to supply chains and trade. But, whatever the precise reasons for the move, it does open the way to rapid advancements in crop science. 'Genome editing is the most exciting technology that I have seen in my many years working in crop science,' said Wendy Harwood, head of the John Innes Centre's crop transformation group, in response to the announcement at the time. 'The technology makes it possible to introduce small changes in crop DNA that lead to the characteristics we need such as disease resistance, better nutritional quality or more resilience to climate extremes.'

With this law change the UK joins the US, Canada, Argentina, India and many more countries in opening itself up to more efficient ways of helping to future-proof food supplies. A good example is scuba rice, which was developed to withstand flooding better than other grains, and is now grown successfully in south Asian countries such as Bangladesh.[5] Scuba rice and initiatives like it will become ever more important as more countries, including the UK, become increasingly susceptible to flooding and resultant crop damage.

GM could also help to reduce waste. According to a 2021 report,[6] approximately 2.5 billion tonnes of food – 40 per cent of the total amount produced each year globally – goes uneaten. Much of this wastage takes place before the produce has even left the farms, in part due to disease and bad weather, or being rejected on account of cosmetic imperfections. Using Crispr and other editing tools, scientists can tinker with genes to make potatoes more commercially attractive, resulting in fewer of them being thrown away and more of them reaching our kitchen tables.[7]

Ocean genomics

Phillip Cleves is a principal investigator at the Carnegie Institution for Science in Washington, DC. Ever since he was a small child, Cleves has had a fascination with corals. Ask him why and he makes a convincing case. 'They have algae plants that live inside of their tissues and collect sunlight from photosynthesis and feed the animal ... That's like us having a plant that grows in our skin and every time we want to have lunch, we walk outside and we get fed lunch. How awesome is that?'

Growing up around 800 kilometres from the coast in Oklahoma, and later Arkansas, Cleves admits he had 'minimal ocean experience', but a year unhappily spent studying poultry farming in the sticks helped convince him that a move into marine biology would be a life well spent. 'To think that one organism could live inside of another organism? And not only do that, but be so efficient as to allow them to build ecosystems that you can see from space? As a kid, and even as an undergraduate, it was just so alien, so bizarre to me. The scale and the beauty and the craziness of that biological phenomenon was what got me into it,' he says.

Like many geneticists today, Cleves graduated just around the time when genetic sequencing and the quest to find the human genome were coming to the fore in science, and so applying to do a PhD focusing on genetics seemed like a natural step. He was surprised to find that nobody appeared to be actively applying these technologies to marine biology, or at least not publicly at the time. While Cleves doesn't claim to be the first person to take genomics to ocean science, he was certainly one of the first to have the foresight that this might be a field that could benefit from such tools. 'My idea was to go and get a PhD in genetics and genomics and become expert enough in that, and then bring it over to marine biology – a field that when I started had not really had a lot of genomics and genetics inbuilt.'

Cleves got onto a PhD programme in the University of California Berkeley's molecular biology department – which just happened to be where Jennifer Doudna was working on Crispr. 'The wave of Crispr excitement had hit Berkeley, so the obvious thing to do in my mind was to try this genome editing tool with corals, but it was a little harder than I had anticipated.' Cleves needed to collect coral eggs in order to study the genetic information they contain but corals generally spawn their eggs just once a year, in the middle of the night, by the full moon – and the species that Cleves was working with were in Australia. 'So we packed up the whole lab, flew it out to northern Australia, and waited for the corals to spawn.' After several sleepless nights, Cleves and colleagues collected the coral eggs, and after fertilisation, injected them with reagents that would allow them to modify the species' genomes. 'We made the first genetically modified coral with that method,' he says.

The value in genetically modifying something like a coral might not appear obvious at first. But for experts like Cleves, there was value in exploration – especially when we consider how little we understand about the genetic make-up of these marine species. What's more, coral reefs are a major resource for coastal and island communities the world over – they provide food and income through tourism, and they are key to ocean biodiversity. But rising sea temperatures are causing corals to bleach and die at an alarming rate: analysis by the Global Coral Reef Monitoring Network suggests warmer ocean temperatures killed about 14 per cent between 2008 and 2019.[8] Protecting them, therefore, will mean finding ways to counteract the effects of even the most optimistic targets for reducing global temperature rise.

What Cleves's team does is not so different to what cancer researchers do in their studies of tumour tissues: they pull the genome apart in a safe lab environment to see what effect each gene has on the species' existence. Cleves likes to use the analogy of an alien coming to Earth for the first time: 'The alien might see a Volkswagen van, for example, and think, what is that? And so they start doing experiments on it. They might start taking the wheels off the van to see what happens to its function – and it turns out that if you take the wheels off a van, it can't roll down the street. And so the aliens can infer that the wheels are actually important for the movement of that van.' Picking the corals' genome apart in this way can therefore help researchers to better understand how the species function. And if we can understand that, we can better protect them from the damage they face from climate change.

In 2020 Cleves's group had a significant breakthrough: using Crispr, they successfully worked out which specific gene was crucial for corals' heat stress management. 'And so we found a gene that's basically a master regulator of whether or not corals will survive heat,' he says. 'Now we're curious about whether or not corals are using that gene to evolve naturally to adapt to the rise in sea temperatures. Now that we know this gene has this function, we can ask, are there corals that have natural variation in this gene that might be the ones that are going to survive the next ten, twenty, thirty, or even 100 years of climate change?' Theoretically, scientists could utilise traits like these to genetically modify coral species to become more resilient to climate change, faster.

In La Jolla, California, researchers at the Salk Institute's Plant Molecular and Cellular Biology Laboratory are doing just that on dry land. Their Harnessing Plans initiative combines traditional horticulture and Crispr-based gene editing to design plants with deeper and stronger root systems capable of absorbing higher levels of carbon dioxide – a process called carbon sequestration. Not only are these plants more resilient, their stronger root systems prevent erosion, another byproduct of warming temperatures, which will make soil more healthy and boost production. At scale, the plants could absorb enough carbon to slow down climate change altogether – a win-win for plants and the planet.

Marine biologists are not quite there yet with corals – there are probably more than 100 genes in total that relate to corals' ability to cope with climate change, and Cleves for one is hesitant to interfere prematurely, 'because we just don't know the long-term impact that might have'. In the meantime, their priority is to find a huge diversity of genes that may have been responsible for helping corals survive this far through natural selection and protect those qualities. Scientists can help corals to speed up their evolutionary process, for instance through selective breeding, in the hope that the whole population can evolve in time as the environment around them changes.

It's unlikely there will ever be a single magic fix for the coral reef deprivation happening across the world, and Cleves stresses that action against sea temperature rise is still vital and should come from governments. But it's hoped that by deepening our understanding of coral resilience, both conservation efforts and policy making can be improved. 'Coral reefs support the livelihoods of billions of people, they're one of the most biodiverse ecosystems on the planet,' he says. 'Studying this system and using genomics and genetics in ecologically important systems like corals will allow us to better predict what's going to happen if we don't do anything in terms of reducing CO_2 emissions.'

Climate forensics

Environmental DNA surveillance is another way in which genomics is taking climate change studies to a whole other level. Advances in sequencing tools mean that instead of just taking a sample to map out the genome of one species at a time, researchers are able to sample a whole ecosystem in one go. An extract of soil or sea water, for example, could reveal the footprint of millions of microscopic inhabitants living there.

It's a method used by Stephen Palumbi, professor in marine sciences at Stanford University's Hopkins Marine Station, for monitoring kelp forests near his home on the California coast. These underwater seaweed forests grow to 30 metres high and provide a home to around 800 marine species including rare fish, sea otters and even baby grey whales. But they also play a fundamental role in regulating their surrounding climate, by absorbing and storing large amounts of carbon dioxide. In doing so, they regulate acidity levels and boost oxygen levels in the ocean which, in turn, all helps to support marine life.

Like coral reefs, kelp forests are under threat. A sudden rise in sea temperatures from 2014 to 2016 killed off more than 96 per cent of bull kelp along a 350-kilometre stretch of California's north coast,[9] according to research by the University of California's Coastal and Marine Sciences Institute. On top of this, kelp forests along the coast of southern California are estimated to have reduced by 75 per cent in the past century due to pollution and fishing.

Palumbi's team are interested in monitoring species that live there in addition to the kelp, which they do in one of two ways. 'We can take a water sample out of the forest,' says Palumbi, 'or we go down there and find small cobbles that sit at the bottom of it. They're covered in algae and little critters running around them, and worms, and crabs, and snails, and all that. We wipe off the slimy surface of the cobble, extract all the DNA from that and then you have this complicated mixture of DNAs of plants and animals, bacteria, viruses, all that stuff.'

The genomic trick involves taking the mixture of DNA and sorting it out to determine which genes belong to which species. This can be done quite efficiently by picking out one reliable gene that all plants and bacteria share in common and sequencing copies of that gene to find comparisons. It sounds complicated, but it can be done in one step thanks to automation, and the results are ready in an hour and a half. 'You can think of it like the genome of the whole ecological community,' says Palumbi.

Environmental DNA sequencing is a useful tool in monitoring and recording species and allows researchers to keep an eye on kelp forest decline. 'It's looking at the biodiversity, the dominant invasive species, species that have switched from one genotype [an organism's complete set of genetic material] to the other. It's a snapshot of the genetics of not just a single species but hundreds of species at the same time, so it's a pretty cost-effective way of getting lots of data.' In a way, environmental genome sequencing is the cheaper, faster modern-day equivalent of Charles Darwin's travels across the world making sketches of the species he found. 'This is what genomics is all about. Every couple of years, the amount of data that we have access to doubles,' says Palumbi. 'It's something we could never have done five or ten years ago, and now it's pretty simple.'

Speaking to Palumbi is an uplifting experience because he frequently drops in the phrase 'after this crisis', referring to life after climate change. 'The ocean will survive and we will help it, that's our job,' he believes. Palumbi's role, and that of his fellow environmental genomicists, he believes, is to conserve marine species to help us buy more time for navigating climate change: 'I can't personally convince the leaders of the world to reduce CO_2 the way they should, there's other parts of our society that are doing that. But I can add my voice by saving as much as possible for the next century, so that when things do start getting better – and they will – there's something left to grow back from, that there are coral reefs in the world after this crisis.'

Sometimes Palumbi's work takes him to a tiny atoll in the Pacific Ocean off the Micronesian islands of Palau. 'No one lives there, there's no one else around, and even so I go there and I can see that climate change is affecting that island. You can see that in the corals there. But you can also get a sense that, in a hundred years, this atoll will still be here, it will just be different. And a thousand years after that, it will still be here, but it will be better.'

5

The future of us

At eighteen months old, Aurea Yenmai Smigrodzki is inquisitive like any other toddler. She likes peanut butter, the beach and mobile phones – or any toys that look like phones. She likes to copy her mum and dad, Thuy and Rafal, when they are using theirs. Aurea doesn't know it yet, but her birth was very special: she is the world's first PGT-P baby, meaning she is statistically less likely than the rest of us to develop a genetic disease or disorder through her life.

PGT-P stands for 'preimplantation genetic testing for polygenic disorders'. It is conducted in conjunction with IVF and allows prospective parents to actively select which of their own embryos to take, based on the strength of its genes. Rafal and Thuy were given the genetic profiles of five prospective embryos, and Aurea's was the strongest candidate, because her embryo had the fewest recognisable genetic mutations that could go on to cause disease. 'It was really a no-brainer,' says Rafal of the choice he and Thuy made to undergo the genetic screening process. 'If you can do something good for your child, you want to do it, right? That's why people take prenatal vitamins.'

All parents want their children to be healthy, but lots have reason to fear passing on something harmful. Our genes can predispose us to developing all kinds of diseases: diabetes, heart disease, cancers and many more. With this in mind, one could be forgiven for assuming that Rafal or Thuy carried some inheritable condition and they wanted to break the chain. But the reality, Rafal admits, is that he 'simply knew that PGT-P existed', and so he decided to give it a try.

Rafal is a neurologist and has an interest in pioneering technologies, referring to himself as a 'techno-optimist'. He has even signed up to have his brain cryogenically stored when he dies, in the belief it will one day be resurrected, thoughts and spirit intact. In his eyes, genetic screening of embryos is nothing crazy or even special, it is simply the natural next step for humans to take. 'It's like the first time someone ever made a phone call – sure, it was a unique moment, but really it was just the beginning of something that now everybody does,' Rafal muses. 'In ten years' time, this kind of polygenic testing will be completely non-controversial. People will be doing it as a matter of course.'

Thuy and Rafal screened their embryos through Genomic Prediction, the first of a couple of biotech firms in the US to open up genetic screening services to prospective parents. Taking DNA samples from the embryo cells alongside genetic sequences from both parents, analysts are able to draw up a set of markers from which they can construct a full genetic picture of the embryo. This effectively fast-forwards its development process to create a projection of what level of health a child born with those genes might enjoy. To help their clients put this data into context, each embryo is given a health score based on the existing mutations in its genes which could potentially one day be life limiting, and the would-be parents are shown how that score compares against the population average. The ranking takes into account the severity of conditions, if shown, as well as the ethnicity of the embryo, since this can also have an impact on disease incidence.

Aurea is the product of that ranking: she was the top-rated embryo out of Thuy and Rafal's IVF collection and the cells they chose to give the best possible chance at living a long, disease-free life. When Aurea is older, she will have access to the full set of embryonic screening data shared with her parents. She will probably have her own genome sequenced, too – Rafal has already purchased a home testing kit for her – and use that information to guide her approach to health and lifestyle through her life. 'I hope she will be glad for it,' says Rafal.

'People ask me if I'm trying to play God in choosing to do this,' Rafal adds, anticipating the next big question. He believes that 'genetic selection is not playing God, it's working as a mechanic on molecular machines that sometimes break and need to be fixed'. Of course, good genes are by no means a guarantee for a long and healthy life, and carrying an abnormality or even living with a hereditary disease does not always equate to a poorer quality of life. Rafal does not for a moment believe that passing on unhealthy genes makes someone a bad parent, either. But he is unequivocal in his belief that he has done the best thing for his child by giving her the best odds against genetic disease. 'As parents, we act as the health champions of our children, and it makes sense to treat genes not as mysterious determinants of identity, but something that you know is there and is important; these are the same principles I apply in trying to take good care of my own health. What matters', he continues, 'is that the process was successful, my child was born healthy, and she is happy.'

Genetic risk factors explained

Polygenic risk scores (also referred to as a genetic risk score) were used by Genomic Prediction in the case of Rafal, Thuy and Aurea to indicate the likelihood of gene mutations among multiple embryos. They are also a marker used in other areas of biology to determine roughly how one organism's genetic health compares to another's. For example, polygenic scores are commonly used in animal and plant breeding to improve the chances of having healthy and resilient livestock and crops. The 'score' is calculated from the number of variations found in each organism's genome that relate to a particular disease (therefore increasing the risk of developing it). This is compared against a reference database compiled from large-scale population studies in order to provide a relative indicator of how likely that organism is to develop a disease relative to the average.

There are no guarantees in using this process: it can only be used as a forecast, because the score only compares to an average organism rather than testing for genetic links to disease in each individual. Neither does it take into consideration environmental factors. For example, a 21-year-old and a 99-year-old could have the same polygenic risk score if their genes predispose them to having coronary heart disease, but the score doesn't account for where they are in their lifespan or when they might present with the disease. So, the indicators are limited, but they can show with accuracy what common genetic conditions a person or organism might be carrying – which is relevant to parents selecting one embryo out of several.

Embryonic selection itself is nothing new. For around three decades, IVF clinicians have taken sperm and egg samples to grow into several embryos at once, before choosing the most promising-looking one for implantation in the uterus. Clinics already tend to screen against chromosomal abnormalities such as Down's syndrome, but until recently the only other indicator they had to go by was the way one group of cells looked against the other – the selection was more or less arbitrary.

Companies such as Genomic Prediction are taking this process much further, giving parents the power to select the embryo they believe to have the best fighting chance of survival both in the womb and out in the world. At the time of writing, Genomic Prediction works with around 200 IVF clinics across six continents. For company co-founder Stephen Hsu, who we met in Chapter 1, the idea behind preconception screening was no eureka moment, but something he and his peers developed gradually. 'We kept pursuing the possibilities from a purely scientific interest,' he says. Over time sequencing has become cheaper and more accessible, and the bank of genetic data has become ever greater, which has provided the opportunity to easily apply machine learning programmes to seek out patterns, Hsu explains. 'You can have typically millions of people in one dataset, with exact measurements of certain things about them – for instance how tall they are or whether they have diabetes – what we call phenotypes. And so it's relatively straightforward to use AI to build genetic predictors of traits ranging from very simple ones which are only determined by a few genes, or a few different locations in the genome, to the really complicated ones.' As Hsu indicates, the crucial difference with this technology is that it's not just single mutations like cystic fibrosis or sickle cell anaemia that the service makes its calculations on. The conditions embryos are screened for can be extremely complicated, involving thousands of genetic variants across different parts of the genome.

In late 2017, Hsu and his colleagues published a paper demonstrating how, using genomic data at scale, scientists could predict someone's height to within an inch (2.5 centimetres) of accuracy using just their DNA. The research group later used the same method to build genomic predictors for complex diseases such as hypothyroidism, Types 1 and 2 diabetes, breast cancer, prostate cancer, testicular cancer, gallstones, glaucoma, gout, atrial fibrillation, high cholesterol, asthma, basal cell carcinoma, malignant melanoma and heart attacks.

This did not come without controversy. In fact, by mid-2020, the outrage among graduate students at Michigan State University was loud enough to force Hsu out of his position as vice-president at the institution. Hsu believes that the opposition people felt to Genomic Prediction in the beginning was largely because people feel uneasy about the fact that genetics can seal our fate – that unfavourable traits can't always be amended through hard work and determination. 'People don't want to believe that there's some degree of hardwiring that can't be overcome by good habits or good education,' he says. 'But the fear is misplaced: the ability to detect single gene mutations has been around for some time and nobody considers that ethically questionable, right? It's just that now we can do it with more precision.'

Studies by Genomic Prediction show that children born through the service have a 46 per cent lower risk of heart attack, 42 per cent less chance of getting Type

2 diabetes, 15 per cent reduction in risk of breast cancer and 34 per cent lower risk of schizophrenia.[1] 'Using genomic predictors we can easily find people who are at ten times the normal risk. We can easily find people who are ten times below normal risk. And that's a huge piece of progress,' says Hsu.

Like Rafal Smigrodzki, Hsu is confident that public disapproval will ease, and that one day soon embryonic selection against inheritable diseases will be considered the norm. In his opinion, 'we shouldn't only use artificial ways to reproduce, but we should make use of the tools we have for IVF to ensure we have the best chance of making healthy babies'.

It's not just Genomic Prediction that is in the market. Other businesses are now offering screening services aimed at prospective parents. One, MyOme, is conducting trials with doctors and IVF patients, the results of which will determine their plans to open to clients. Orchid, a San Francisco-based start-up, launched its waiting list for at-home DNA testing kits in spring 2021 aimed at couples who are looking to have children. The service promises a report detailing risks for any future children and separate male and female partner reports.

Improving the fertility process

One aspect of Genomic Prediction's work that few would criticise, and shouldn't be overlooked, is screening to improve the chances of a healthy pregnancy. In screening for healthy embryos, clinicians are also reducing the chances of complications during pregnancy which can result from genetic factors, and helping couples to save time, money and a lot of heartache. But advancements in genomics are facilitating other ways of improving the conception process for couples, too.

In 2013, a biotech company called Natera became the first to develop a panoramic blood test to screen for foetal abnormalities from as early as nine weeks into pregnancy.

The non-invasive prenatal test (NIPT) replaces traditional chorionic villus sampling (CVS) – an unpleasant test involving a large needle to extract cells from the placenta – in screening for chromosomal conditions such as Down's syndrome, Edwards' syndrome or Patau's syndrome.[2] CVS was effective in detecting abnormalities, but increased the risk of miscarriage by around 1 per cent; NIPT, being non-invasive, is much safer.[3]

Like so many life-changing inventions, the first NIPT came about because of a personal experience. Natera's co-founder and chair Matthew Rabinowitz explains that, in 2003, his sister gave birth to a son who had severe chromosomal abnormalities. Tragically, the baby died at six days old, an experience Rabinowitz says was 'overwhelmingly upsetting. I'm an engineer and my nature is to solve problems. And this was a situation that was just not fixable.' The family felt blindsided: 'He was born in a top hospital in Boston, and yet they didn't realise that he had these problems until he was born,' says Rabinowitz. 'I thought, How could this happen in the twenty-first century? There's got to be a way to improve on these technologies.'

At the time, Rabinowitz was a professor in aeronautics and astronautics at Stanford University, and dealt with the shock and grief in the best way he knew how: by throwing himself into new research. His team began work on genetic techniques to look at tiny amounts of DNA that could extract 'a much more powerful signal' of the health and development of a foetus. Rabinowitz submitted the resulting data to the National Institutes of Health (the US's publicly funded medical research agency), which awarded him a start-up business grant – and the rest is history.

NIPT works by sequencing small fragments of foetal DNA (cfDNA) that can be detected in the mother's blood. The cfDNA is isolated and examined to detect aneuploidies – where an abnormal number of chromosomes is present in each cell – but also, if the doctor requests, some of the single-gene variations which cause severe genetic diseases. Since biological sex is determined by chromosomes, the blood test can also identify the foetus's sex much earlier than an ultrasound. Different brands of NIPT have since been developed and rolled out across the world in both public and private healthcare. 'It's completely changed the way people all over the world manage pregnancy,' says Rabinowitz.

Tracking the biological clock

One of the main reasons couples undergo IVF treatment is because one or both of them has a fertility issue. This is becoming increasingly common as, particularly in the Western world, we're having families later. The data shows that the average age at which women have their first child has increased steadily in line with equality. Figures from the Office for National Statistics show that in England and Wales, the average age for first-time motherhood in 2020 was 29.1, up from 23.7 in 1970,[4] and a similar pattern is true of other wealthy countries.[5] In this sense biology has dealt women a particularly unfair hand in that our fertility levels famously drop off at speed from our mid-thirties onwards – regardless of how emotionally or financially ready we may be for motherhood by this time. But individual fertility is unpredictable because not everyone experiences fertility decline at the same rate – and it's more or less impossible to know which side of luck a person falls on unless they choose to get their fertility tested at cost in a private clinic. But even here, genomics might have a trick up its sleeve.

Research led by John Perry, a geneticist at the University of Cambridge, is tentatively offering some clues into predictive factors for fertility decline in women – and the findings could even help some of them to avoid conception problems altogether. Along with colleagues at the universities of Exeter and Copenhagen, Perry scanned the genes of more than 200,000 women and found nearly 300 genetic signals that could help identify why some are predisposed to early menopause, along with the associated health consequences, and whether these clues might be manipulated to improve chances of conception.[6]

Two genes named CHEK1 and CHEK2 were found to be a key factor in understanding the difference between women who go through early menopause (aged forty-five or younger) and those who do not. Perry and his team conducted a study of mice in which the CHEK2 gene was shut down and this resulted in offspring which were able to reproduce at a later average age. Similarly, when CHEK1 was overexpressed mice were born with larger numbers of eggs to begin with, meaning their reproductive lifespan was also extended.

The discovery opens up the possibility for earlier detection of fertility drop-off before it happens. 'One of the motivations of the work was that, at the moment, we have very limited ability to identify who in the population will suffer from infertility,' says Perry. A twenty-year-old woman with no particular family history of any problems with fertility, for example, might wonder at what age she might stop being able to conceive children. 'At the moment, there's absolutely nothing that we could measure that would give you that answer. The best we can do is measure the level of certain hormones in women.' Doctors can measure for a hormone called anti-mülleria, which gives a short-term read-out of the egg reserves left in a person's ovaries. But the insight is limited: doctors can't use it to predict several years into the future, and the test can only determine whether the reserves are well stocked or low at that moment. And if the reserves are low, the likelihood is that the women taking the test would already be having problems conceiving.

Just as is the case with other diseases, infertility can be caused by a myriad of things. In rare cases, a person might have a single causative gene mutation that negatively impacts sperm production or sex hormone regulation or other factors that are crucial for functioning fertility. Others might have an unlucky accumulation of lots of more subtle genetic factors which, added together with environmental risk factors such as stress, obesity and smoking, might make conceiving difficult. In these cases, fertility can be harder to predict, but as screening becomes more sophisticated, experts are hopeful that they will one day be able to paint a fairly reliable picture.

'We're really motivated to find long-term predictive tests, because what's hard for couples is the uncertainty of it all,' says Perry. Ultimately, he imagines that these fertility-related genetic factors might be screened for at birth, so that everyone can go through life armed with more knowledge of what their natural fertility window might be in years to come. 'Obviously this prediction might not be perfect, and someone could still face fertility issues as a result of other non-genetic factors. But they'd be better equipped to make more informed decisions rather than just rolling the dice,' says Perry. The benefits of doing this would extend far beyond emotional reassurances, too – it could save individuals and public health services a lot of money in IVF costs simply because couples will have a clearer idea of when they need to start trying for a baby if they want one.

The research may still be in its early stages, but it just might lead to more genetic secrets being revealed, which will in turn allow us to bend the rules of biology to fit in with modern life and all its pressures. The very fact these new avenues are being explored will help to bring about wider positive changes in equality, too. An open letter published in *Nature* by a group of leading advisers into global women's health in 2020 highlights the significant gender gap in data: 'Women's health issues, and their preferences, are simply under-studied and under-funded, and unmet needs are ignored and misunderstood by those who could work to address these issues,' the authors argued. It was only in 1993 that the US National Institutes of Health introduced a rule to ensure women were included in human studies as well as men. There is currently no such rule in the UK.

Women's healthcare has lagged behind in terms of research and advancements, but a greater understanding of the genetic factors relating to fertility and the menopause could bring positive changes to the medical care available to women as well as our societal attitudes towards research.

Genetic screening: the way to go?

'In our lifetime, it's entirely likely that every time a baby is born its genome will be sequenced, and its genetic predisposition for every known disease will be evaluated from birth,' says Perry. And he is not alone in thinking this. Just about every person interviewed for this book agrees with him – and that's because genomics has the potential to bring about massive advancements in healthcare. Routine genome screening with disease predictor scores could help public health systems to move closer towards a culture of prevention rather than cure, which is 'the ultimate goal of public health', says Perry. Instead of chasing treatments after problems have occurred, screening could help us to pre-empt them, which is better for the taxpayer and much better for the patient. We know by now that the technology exists for that to happen, but how long it will take for society to accept such a move is less easily determined.

One of the big challenges in public acceptance of this technology is in better communicating the science of genomics and the consequences of knowing more about our genetic make-up. Perry's wider research projects are based around the determination of genes which predispose people to different health risks, but also the impact of metabolism on different conditions such as diabetes. Metabolism, says Perry, plays an equally important role in our general health and wellbeing. With this in mind, he is keen to get the message across to people that our genes do not guarantee whether we will have certain traits or diseases. For example, having a high genetic predisposition to diabetes doesn't automatically mean that a person will go on to get the disease – other lifestyle factors play a role. Consider that people who win Nobel Prizes tend not to be young. But most people who are old don't have Nobel Prizes – it's the same sort of logic for these genetic risk factors. 'We develop most common diseases due to a critical mass of environmental and genetic risk factors. Any individual risk factor won't necessarily push you into that category and a healthy lifestyle might help offset the risk from genetic predisposition,' Perry explains.

Another big challenge to the rollout of genetic screening is that, as humans, our perception of risk is not great. This is something that has been shown through dozens of socioeconomic studies, but also from real-life examples of our ability to ignore risks that are seemingly invisible to us. Take for instance the Covid-19 pandemic: even though the statistics on rising infections were clear, some people still avoided following social-distancing rules and refused life-saving vaccines. For this reason, it can be very difficult to change societal behaviours. Common health messages – lose weight to avoid Type 2 diabetes, cut down drinking to avoid heart disease and liver failure, stay at home during the pandemic – go against the grain of the freedoms so many of us are lucky enough to enjoy in life. This means they are often taken seriously only after the event, when we can see the personal implications of ignoring such advice: after a heart attack, after a cancer diagnosis, after losing a loved one.

'If people are hesitant to even take a vaccine or something that has a very clear and obvious benefit for them, imagine how a challenging conversation about sequencing the DNA of their children might go down,' says Perry. In all likelihood, just like with vaccines, society will become split between those who do consent to genetic screening for themselves and their family and those who do not. 'The science is only just getting to a place where we can see this on the horizon and the clinical, translational studies will need to catch up with it, to help inform the resulting policies,' Perry adds.

Wanting to learn more about the way our bodies work, and where their flaws lie, is a completely natural thing to geneticists and indeed many other people with a scientific background or interest. Figures from the DNA home testing kit company 23andMe show some 12 million people have had their genome sequenced by the service so far, but more than a quarter opt out of discovering potential health risks as outlined in their genome.[7] Which goes to show that what might seem obvious to one person goes against the path of logic for another. There's something to be said for this attitude, too: what would the value to someone be in knowing they have a chance of developing something like Alzheimer's or Huntington's disease while there is still no cure? There is no logic in burdening people with the knowledge only to leave them anticipating their own decline.

At Mariya Moosajee's eye clinic, there are strict ethical rules on the information that can and can't be shared from genetic screening tests. 'For example, sometimes a parent has a genetic condition, and they are worried about having passed it down to their child. If that child comes to my clinic and they are asymptomatic – they've got no signs of any eye disease – I cannot do a genetic test on them until they are sixteen and can legally make the decision for themselves,' she explains. 'It's because there are too many possible implications. If you're doing genetic tests and labelling someone with a condition when they feel otherwise completely normal, it can stir up a lot of anxiety, fear and mental health issues.' It's for this reason that Moosajee believes it's unlikely that patients of any age will be told if their genome sequence has revealed a predisposition to dementia, for example: 'It could have a massive negative impact on your life that outweighs any benefits in knowing.'

Mass rollout of genetic screening from birth is not likely to happen just yet, according to Moosajee, because there's 'a lot of ethics to move forward with and to get over'. 'But I do believe that in the future, when we do the standard heel prick test, that drop of blood will go on to be whole-genome sequenced. Your genome map will essentially sit with you for your lifetime and it will be used as a reference point at various times in your life. But, it will always have to come down to consent.'

Genomic studies have shown that the potential for better medical care, even before birth, is great. Similarly, scientists are tapping into new clues about the way our bodies play out genetic factors all the time, and this is being applied to public guidance on personal health, fitness and general wellbeing. As far as self-care and education go, it's all good news: with better understanding we can be better armed to look after our bodies and stay healthy. But the future of medical ethics and consent is much more fraught, and the rules are still very much being written.

Drawing the line

The debate becomes particularly murky once one moves beyond disease control and eradication and into the realm of creating the 'ideal' human. Right now, not only is it possible for Stephen Hsu's team at Genomic Prediction to determine whether or not an embryo carries a risk of diseases, they can also allow parents to select for much, much more including, for example, skin colour, eye colour, height, even intelligence. Hsu says he won't, and doesn't, do any of those things. 'It's not an option. We draw the line at anything that's not disease-related or directly impacting the health of the embryo – we're not giving any other information to the couple.'

However, even for Hsu, some of the lines are blurred. Take intelligence, for example. While there is no one 'smart' gene, research suggests there are many genetic factors relating to a person's intelligence. In one study comparing DNA variants from more than 240,000 people, researchers identified 583 genes linked to intelligence and 187 areas in the human genome that are associated with thinking skills.[8] While it's still difficult to predict how clever someone might be using just their genomic data – environment plays a role, too – the same study suggests that around 50 to 80 per cent of variation in general intelligence between people comes down to genetics.

While Hsu does not necessarily endorse screening for intelligence, he is interested in the emotional response people often have to the idea. 'If your neighbour's kid is intelligent, it doesn't preclude yours from being intelligent, too,' he says. 'It's what we'd call a non-rival good.' The argument goes that, unlike financial wealth or other factors of privilege, there's not a limited amount of cleverness available for the taking, so we should feel comfortable with the idea of everyone inheriting intelligent traits. Rafal Smigrodzki, Aurea's father, appears to agree: 'It makes sense that Genomic Prediction don't allow for [selecting for intelligence] yet,' he argues, 'because there is already a minor storm over the technology itself and you don't want to endanger the right for the technology to exist by inviting more controversy. But the fact is, intelligence is one of the key determinants of health.'

The case for such interventions involves various strands. As Rafal says, intelligence is not a finite resource. It's also true that there's a link between intelligence and better physical health, not least because more qualified people earn better money and so are more likely (in the US, certainly) to be able to afford good healthcare. Then there's Rafal's argument that 'smart people have the power to do so much more for their society. It's all about what you do with that power.'

Many would argue that such theories are pitted with flaws and dangers. One view, for example, that there's a particular type of 'intelligence' that is simultaneously 'best' and that can be clearly defined and engineered for is a deeply problematic one. Human achievement comes in many different forms that cannot be reduced to a single notion of intelligence: practical intelligence, academic intelligence, emotional intelligence all have a part to play in the shaping of the world, and we need people with all types of skills and interests to make it thrive. In any case, there is much more to intelligence than straight genetics. Background and environment also have crucial roles to play.

It's not difficult to see, then, that a field with its roots in scientific research has much broader ethical and philosophical implications.

6

Ethics and responsibility

On 26 October 1966, the American biophysicist Robert Sinsheimer addressed a packed-out theatre at the California Institute of Technology.[1] 'We will surely come to a time', he said, 'when man will have the power to alter, specifically and consciously, his very genes. This will be a new event in the universe. The prospect to me is awesome in its potential for deliverance, or equally, in its power for disaster.'[2]

The underlying sexism of his terminology aside, Sinsheimer was nearly forty years ahead of his time, and the words spoken at that event still ring true. Genomics, and the possibilities opened up by gene editing, is a subject that continues to terrify as much as it fascinates. There are big questions to answer: where, potentially, could genomic technologies take us? Where will future technologies sit alongside our ethical and moral boundaries? To paraphrase Dr Malcolm in *Jurassic Park*, just because we have the power to do something, does it always mean we should?

A quick scan of popular culture over the past few decades, even centuries, attests to the power these questions hold over us, and their potential to inspire dystopian thoughts. From the publication of Mary Shelley's *Frankenstein* in 1818, to Aldous Huxley's *Brave New World* in 1932, to more contemporary science fiction books, TV series and films like *Gattaca* (1997) and *Black Mirror* (2011–19), and all kinds of low-budget releases in between and since – the manipulation of genetics, the very fabric of what makes us who we are, has a huge influence in our understanding of identity and right and wrong, and has frequently come up in public debates and conversations when genomics is mentioned.

Genetic engineering has been presented by creatives and thinkers as a brave new world of exciting possibilities, but it is just as often a trope offered as a warning against the perils of messing with nature. Speaking about his own fiction writing on the subject, the Nobel Prize-winning author Kazuo Ishiguro has commented, 'I worry about the meritocracy of the world becoming a biological aspect as well.'

Where public fear exists around genetic sequencing, it is no doubt influenced by our troubled history with eugenics, particularly in Europe. Originally from Germany, Stephan Beck, director of the Personal Genome Project in the UK, is very familiar with the complex feelings societies hold about the subject. He moved to the UK partly in order to avoid some of the backlash around genetic sequencing he had witnessed in Konstanz in Germany in the 1970s. 'We were just beginning to sequence yeast, but there was huge public outrage,' he recalls. 'People were protesting in the streets about it. And this was just with yeast – imagine if we had been publishing the results of animal or human studies. The message was that we've gone down this road in this country before, and it was done wrong – so we just shouldn't touch it.'

Fifty years on, the subject is still a sensitive one, and there is much left to address that modern laws are yet to catch up with. But Beck senses a change in public attitudes. 'The UK public is particularly well educated and knowledgeable about genomics, I believe, partly because organisations like the Wellcome Trust and various other big players have really played a role in reaching out to educate the public about it. In the US and parts of Asia, too, I think people are much more relaxed about this subject. It's taken Europe a long time to catch up.'

The role of ethical counselling

Continued efforts to communicate the facts on genomics to the general public will be crucial if the science is to be widely accepted and utilised in the best way. This is especially the case where patients are involved. The UK's National Health Service, for example, is seeking to further public understanding through genetic counselling. Jonathan Roberts is a researcher in the Society and Ethics Research Group at the Wellcome Genome Campus in Cambridgeshire, but he has many different hats. He also works as an NHS genetic counsellor at Addenbrooke's Hospital in Cambridge, speaking to patients in the prenatal clinic one on one about family history and the possible implications of inherited conditions.

There are two main reasons people come in to see Roberts at the prenatal clinic. One is if there's a known genetic condition in the family and a couple or a woman are trying to choose whether or not to have genetic testing during or ahead of pregnancy. Another is if a woman is already pregnant and a routine scan has brought up an abnormality that's suggestive of a genetic condition. Separately, Roberts will sometimes have an input on decision making in his colleagues' diagnostic cases, providing an ethical point of view in how to go about treating cancer for example, 'especially if there's also particular need in these cases for supporting colleagues with the emotional dimensions of the case'.

When a couple or pregnant mother has been referred for a genetic screening, a detailed investigation must first take place. The patient and partner receive a family history form to fill in, which Roberts and his colleagues will assess before making a judgement on whether DNA sequencing is needed. 'In that situation a big part of our role is to try and confirm information to get more of an accurate picture of the family history,' he explains. The problem here is that families often don't communicate well between themselves when it comes to matters of health, or, as is quite common, a story of a relative's cancer or other illness gets warped across different conversations and generations.

'Say someone's referred to us after a diagnosis of breast cancer – they might tell us they had a relative with ovarian cancer, which can be indicative that they're carrying a particular mutation which predisposes them to both of these cancers. Quite often we have to try to confirm that it was ovarian, because often cancers get misreported in families, so it might be that the relative actually had, say, cervical cancer – and that changes the whole risk assessment because it's not an inherited cancer.'

The genetics counselling team can request pathology reports to firm up details here – which brings added complications as it sometimes requires asking other family members to consent to them accessing their medical records. 'Sometimes my job is ringing people up and sort of saying to someone . . . "A consent form hasn't been sent back to us from your aunt. What's going on, is there a rift there?" And trying to explain to people why we need that information. It's a lot of background work just trying to get as accurate a picture as we can about the family history so that we can make a reasonable risk assessment.'

The work doesn't stop there: Addenbrooke's has an entire family forensics team called the Patient Information and Collation Service, whose job it is to chase and track down gaps in family histories. 'The family history form might say that someone's father had colon cancer in Norwich in 1987, for example, and there's different ways to confirm that. You can track them down through a cancer registry at the hospital there, or look at births and deaths records. If, for example, someone is reported to have been diagnosed with ovarian cancer in their thirties in 1985 but they're still alive now – we can look at the odds of that and deduce that it probably wasn't ovarian cancer at all.'

The investigation team builds a picture, and some of their cases will tick enough boxes for concern so that the patient goes on to get their genome sequenced by the hospital to help them determine the likelihood of passing unhealthy genes down. Others don't. The reasons for this largely come down to resources – Addenbrooke's is a publicly funded hospital and the cost and time taken to sequence every genome would be too much. But this is changing, especially where cancer is concerned. Since 2021, whole-genome sequencing has been rolled out for use on NHS cancer patients more widely, and the fact the technology is becoming faster and cheaper helps. The NHS Genomic Medicine Service is already available for patients with suspected cancer and other serious conditions, and it's hoped that the service will help to reduce waiting times and adverse drug reactions. It could also help to determine the health risks of having children, or what living with a health condition during pregnancy could mean for the child.

Public discussion is already taking place around whether or not genetic screening services could be rolled out more widely in the future – perhaps as a service offered to all couples who want to know the risks before conceiving. 'It's a really interesting question,' says Roberts. He feels confident, however, that family background investigation won't be replaced by mass availability of sequencing entirely, 'because the genome data on its own actually isn't that useful. As genomic sequencing becomes more and more part of our healthcare, we're still going to need all of that other information to guide the analysis and actually make sense of what it means, both from an individual context and for wider healthcare planning purposes,' he says.

While the NHS has clear guidelines on what does and does not warrant genetic screening, private companies do not – and it's entirely likely that patients who are denied genome sequencing through their doctor will go about it another way instead. Roberts has concerns about the commercialisation of these services. 'The challenge is going to be regulating private companies, especially around this idea of polygenic risk scores,' he warns. 'Deep DNA is an easily sellable cultural idea, particularly where anxious couples are being targeted. But there is a huge amount of uncertainty, and actually the predictive value of genetics is not that high.'

The problem he foresees is that at-home commercial DNA tests work well on a whole-population level, but the individual predictive power – of whether or not someone's child might develop schizophrenia, for example – might not be so accurate, because it's based on data averages. 'Polygenic scores are impacted by social data as well as biological data,' he says, referring to the fact that they draw from data which is not accurately representative of the different ethnicities or social backgrounds of the general population, 'so using them in a prenatal setting is problematic'.

UK health advisers are working hard to improve public understanding of genomics in this regard. The NHS National Genetics and Genomics Education Centre was established in 2005 following a Department of Health paper, 'Our Inheritance, Our Future: Realising the Potential of Genetics in the NHS', which was commissioned shortly after the publication of the first human genome sequence. State-sponsored websites such as tellingstories.nhs.uk seek to educate people on what sequencing does through real examples. It's hoped that gradually, genomic sequencing won't be seen as something mysterious and frightening but as another tool in public healthcare.

There is also work underway to expand and formalise the information available to the public on genome editing technologies. More recently, the Wellcome Society and

Ethics Research Group that Roberts is part of has joined up with public participation charity Involve UK and Keele University to launch the Citizens' Jury on genome editing in the UK. This jury is planned to take place in the first half of 2022, and is part of a broader vision to run a Global Citizens' Assembly on the same topic – the idea being to keep the public informed and actively part of policy decisions on the future of genome editing technologies. The three UK partners form part of a global consortium of organisations led by the Centre for Deliberative Democracy and Global Governance at the University of Canberra, under which projects are already confirmed in the US, Australia, China, South Africa and across Europe.

Abuse of power in gene editing

In March 2015, a group of leading US biologists wrote an editorial comment in *Nature* magazine titled 'Don't edit the human germline'. Research that experimented with this new area prompted 'grave concerns regarding the ethical and safety implications', they warned. More than this, the therapeutic benefits of modifying human genes at this stage were still 'tenuous', and the potential for public outcry could do a lot to damage valuable research in other areas of genetics going forward. Many scientists seem to agree with them, even (and perhaps especially) the inventors of Crispr itself. Speaking in the 2019 documentary *Human Nature,* Jennifer Doudna spoke of a dream she had some years ago where she came face to face with Adolf Hitler. 'He leaned over and he said, "So tell me all about how Cas9 works." I remember waking up from that dream and I was shaking. I thought, Oh my gosh, what have I done?'

Gene editing in humans takes one of two forms. In somatic gene editing, changes are made to DNA in a whole living person, that is, in the cells in their body. In germline editing, changes are made to the DNA in embryos, sperm or egg cells, where the inheritable genetic code lies. The difference between them is a crucial one: somatic gene editing physically affects only the person it is conducted on, while germline editing results in changes to that person and any descendants who follow.

Germline editing has the potential to change generations therefore, and on a larger scale, our entire species and evolution process. Understandably, it's much more controversial. Unlike targeting cells in children or adult humans, changing the germline is irreversible. Moreover, the technology is simply too new for us to know what the long-term effects of that irreversible change might be. More than forty countries prohibit germline editing in their laws, but there is no one universal rule across continents. In the UK, for example, use of genome editing in embryos is expressly prohibited for reproductive purposes, but researchers can do germline editing on embryos that are discarded after IVF, so long as they are destroyed immediately afterwards. In July 2018, an inquiry into the ethical issues of germline editing in the UK concluded that there was 'no absolute reason not to pursue it' in future, but that appropriate measures must be taken and strengthened by law. 'The implications for society are extensive, profound and long-term,' the inquiry's chair, Karen Yeung, added.

In November 2018, the Chinese biophysicist He Jiankui announced via YouTube that two genetically modified babies, Lulu and Nana, had successfully been born, 'crying into the world as healthy as any other babies'. He had used Crispr technology to modify the DNA of human embryos, targeting a specific gene called CCR5 in an attempt to create babies who are resistant to infection from HIV.[3] The experiment was ripe with controversy, not least on account of it involving eight couples in the project which included male partners who were HIV-positive. But through artificial insemination, the experiment had worked, He confirmed. Speaking at an international conference in Hong Kong two days later, He announced that a third genetically modified baby would soon be born.

Shortly after the news from China emerged, scientists around the world called for a five-year moratorium on gene editing of humans until the technology could be proven to be safe and efficient. At the end of 2019 He was sentenced to three years in prison for violating medical regulations.

Editing our future selves

There is a gene responsible for pain. There are people in the world who can't feel pain because they have a mutation in one single gene, SCN9A, which is responsible for making a protein that transmits the pain signal from the source to the brain. Without that gene, the individual can cut themselves or walk over fire – jump off a bridge, even – and feel no pain. With this knowledge, could researchers get rid of that gene and engineer a world without pain? It's surely a tempting prospect when considering the pain caused by cancer and other diseases, for instance. But there are reasons why we have evolved to feel pain, too – it's the body's way of alerting us not to touch something that could damage us, for example.

The technology to eradicate some diseases such as cystic fibrosis already exists. But whether or not it will be taken up will depend on the decision made collectively by society. Not everyone living with an abnormality or disease would wish it away – to assume they would is patronising and reductive. 'Maybe one day they will be able to get rid of the CF gene in humans completely,' says Jess Spoor, the CF patient whose story features in Chapter 2. 'I am very grateful, however, to have lived with CF as it has shaped the person I am today. It has taught me many things, especially that every moment is precious and I have a great appreciation for life. It has made me stronger and very driven to help others.'

In September 2021, a large-scale genetic study of autism spectrum disorder (ASD) was suspended following criticism that its investigators had failed to properly consult the autism community about the goals of the research. Concerns about the Spectrum 10K study include fears that its data could be misused by other researchers seeking to 'cure' or eliminate ASD entirely. There are many examples like this which disability campaigners have opposed for similar reasons. The fact is, plenty of people with a diagnosed disability lead full and happy lives, contributing to society and making the world a more diverse and open-minded place. It's very difficult to draw a clear line between gene editing for 'good' and 'bad' reasons.

More generally, the issue with any kind of systematised genetic engineering is that it is dangerously reductive. It makes simplistic assumptions about what is desirable and what is undesirable. It makes arbitrary decisions about what humans should be, and it runs the risk of destroying precisely the diversity that has helped humans to thrive.

For all the criticism that Stephen Hsu has taken in the past few years, he is firm in his belief that he is doing this work for the right reasons. 'If I was just trying to make money, I would not pursue the strategy,' he says. 'But I think the correct evolution for this is that society has to understand it and then make decisions – and then we can operate the way society wants to ... Now it's here, there's no going back. You're not going to put the genie back in the bottle.'

Conclusion: Genomics' place in the future

It's difficult to imagine a future in which genomics doesn't feature heavily – in human health, climate and biodiversity studies, and more widely. The clues that genomic technologies have unlocked so far have given experts more promise than ever thought possible. But there are big bridges still to cross.

We need to be wary of the ethical and moral boundaries of gene editing, and honest about the destructive potential of these technologies. If history has taught us anything, it's that the best science is done openly and with transparency, and this extends to the way in which its findings are published and communicated. Projects underway such as the Global Citizens' Assembly[1] on gene editing are still relatively new at the time of writing this book, but will become paramount to the direction of public policy in the years to come. The vast amounts of data being generated through genomics must be well managed, with joined-up policies between countries and industries. It will also need to become more diverse in its approach, building more representative population databases using DNA from different ethnicities. Open access to this data will need to be protected at all costs, to prevent 'ownership' of genomics by commercial stakeholders.[2]

In addition, although genomics could have a very powerful and positive impact on our future as a species and as a planet, we need to be very careful about not overplaying its role. Selling genomics technologies as some magic fix for all societal problems is unrealistic and will only fuel public mistrust on the subject further. While all the researchers interviewed for this book speak enthusiastically and passionately about genomics as a tool for furthering our understanding about the world we live in, many are at the same time cautious not to paint the technology as a miracle solution. There are still plenty of factors that genomics on its own cannot solve or provide answers to. It could be that the technology to do so is just around the corner – or that the answers can be found by doing something different altogether.

There is a school of thought among some scientists that the relatively new field of genomics has overpromised and underdelivered. When the Human Genome Project was launched, huge assurances were made about curing diseases and making us healthier, wealthier and on more of an equal footing than before – but the changes have been incremental rather than dramatic. Sequencing technologies and gene editing tools may not have yet eradicated cancer or ageing, saved the planet or eliminated inequalities. But they're about the best tools in the box that researchers have in solving many of these issues – so long as they are used ethically, responsibly and with empathy.

Acknowledgements

This book wouldn't exist without help from lots of people, some of whom are named in the chapters and others are not – all were kind enough to give me their time and thoughts, so thank you, truly.

Thank you to all of my expert interviewees (in no particular order, and I hope I haven't missed anyone): Julian Parkhill, Sharon Peacock, Stephan Beck, Stephen Hsu, Daniel Peckham, Mariya Moosajee, Jeffrey Beekman, Maarten Geurts, Mathew Garnett, Russ Altman, Charlie Swanton, Iraj Ali, Celia Souque, Alex Cagan, Chris Hunt, Julia Wilson, Sonia Goncalves, Phillip Cleves, Steve Palumbi, Rafal Smigrodzki, Matt Rabinowitz, John Perry, Jonathan Roberts, Sarah Teichmann, Kieron Mitchell, Karen Vousden. All of these people know far more about genomics than me – my job was simply to put their brilliant knowledge down on the page in a way that was accessible. If I've managed to achieve some semblance of that, the credit is all theirs.

Thanks also go to Lee Stern at Achilles Therapeutics and Bridget Baumgartner at Revive & Restore biodiversity mission. Harry Dayantis, Alice Deeley and Kathryn Ingham, my fixers at the Francis Crick Institute, have been life savers over the years, and Emily Mobley helped me to find the right people to speak to at the Sanger Institute. Thank you for being so enthusiastic and helpful from the very beginning.

A special thank you to Patricia Kilpatrick at Cystic Fibrosis Care, and to the brilliant Jess Spoor and Laura Gaytor and their families, for allowing me to share their stories so publicly. You can read more about CF and the support available at www.cysticfibrosiscare.org.uk.

I'm indebted to the team at WIRED, especially Greg Williams and Vicki Turk. Vicki was the first WIRED editor to commission me (my first ever pitch as a freelancer, in fact!) and her willingness to take a chance on a stranger's writing is something I will always remember and be grateful for.

My thanks do not go out to the Nobel Prize winner who refused to talk to me without payment (I didn't), but I will be sure to send them a copy of my other book – it's on the principles of fair science and the importance of open (free) access to research.

Thank you to Anthony Johnson, not only a wise man of science and a dear friend, but an artist extraordinaire who was kind enough to contribute the diagrams in this book.

This book would not have been possible without Nigel Wilcockson, Elena Roberts and Rose Waddilove at Penguin, who helped knock my words into shape.

And last but not least, thank you to Dave and Thuli, who get me out of bed in the mornings, lend me their ears, and offer endless encouragement and jokes through the best and worst times. I'm sure some of my family members still think I've written a book about small beardy men with fishing rods.

Notes

Notes to 1 How to map a life pages 8–29

1 https://www.genome.gov/human-genome-project/Completion-FAQ

2 https://www.forbes.com/sites/matthewherper/2017/02/21/can-craig-venter-cheat-death/?sh=678010be1645

3 https://www.nature.com/articles/d42859-020-00101-9

4 https://www.nature.com/articles/d41586-020-01849-w

5 https://www.yourgenome.org/stories/why-was-there-a-race-to-sequence-the-human-genome

6 For example, https://www.frontiersin.org/journals/genetics

7 https://www.yourgenome.org/facts/what-is-the-illumina-method-of-dna-sequencing

8 https://www.ncbi.nlm.nih.gov/pmc/articles/PMC5293262/

9 https://nanoporetech.com/products/smidgion

10 https://www.nature.com/articles/d41586-020-02765-9

11 Source: Human Nature documentary, Netflix

Notes to 2 The health revolution pages 50–72

1 https://sickle-cell.com/statistics

2 https://news.harvard.edu/gazette/story/2021/06/gene-editing-shows-promise-as-sickle-cell-therapy/

3 https://www.gov.uk/government/news/new-implementation-plan-to-deliver-world-leading-genomic-healthcare

4 https://www.genomicsengland.co.uk/the-uk-has-sequenced-100000-whole-genomes-in-the-nhs/

5 https://genomemedicine.biomedcentral.com/articles/10.1186/s13073-018-0552-3

6 https://www.jax.org/personalized-medicine/precision-medicine-and-you/what-is-crispr#

7 https://www.nature.com/scitable/topicpage/pharmacogenomics-and-personalized-medicine-643/

8 https://www.ncbi.nlm.nih.gov/pmc/articles/PMC1888599/

Notes to 3 Protecting against future threats pages 83–89

1 https://www.theguardian.com/world/2020/nov/23/swale-kent-becomes-england-covid-hotspot-cases-rise

2 https://virological.org/t/preliminary-genomic-characterisation-of-an-emergent-sars-cov-2-lineage-in-the-uk-defined-by-a-novel-set-of-spike-mutations/563
3 https://www.bbc.co.uk/news/health-55413666
4 https://www.gisaid.org/index.php?id=208

Notes to 4 Navigating the challenges of climate change pages 96–109

1 https://allianceforscience.cornell.edu/10-things-everyone-should-know-about-gmos-in-africa/
2 https://www.tandfonline.com/doi/full/10.1080/21645698.2018.1476792
3 https://www.theguardian.com/environment/2015/oct/01/half-of-europe-opts-out-of-new-gm-crop-scheme
4 https://sitn.hms.harvard.edu/flash/2015/from-corgis-to-corn-a-brief-look-at-the-long-history-of-gmo-technology/
5 https://www.nhm.ac.uk/discover/the-future-of-eating-gm-crops.html

6 https://wwf.panda.org/discover/our_focus/food_practice/food_loss_and_waste/driven_to_waste_global_food_loss_on_farms/

7 https://www.wired.co.uk/article/farm-food-waste-genetics

8 https://www.npr.org/2021/10/05/1043372978/global-coral-reef-loss-report-climate-change-warming-oceans

9 https://www.nature.com/articles/s41598-019-51114-y

Notes to 5 The future of us pages 123–140

1 https://www.mdpi.com/2073-4425/11/6/648

2 https://www.genomicseducation.hee.nhs.uk/blog/what-is-nipt/

3 https://www.ncbi.nlm.nih.gov/pmc/articles/PMC3893900/

4 https://www.ons.gov.uk/peoplepopulationandcommunity/birthsdeathsandmarriages/livebirths/datasets/birthsbyparentscharacteristics and https://www.statista.com/statistics/294594/mother-average-age-at-childbirth-england-and-wales-by-child-number/#:~:text=In%202020%20the%20average%20age,%2C%20third%2C%20and%20fourth%20child

5 https://www.oecd.org/els/soc/SF_2_3_Age_mothers_childbirth.pdf

6 https://www.nature.com/articles/s41586-021-03779-7

7 https://blog.23andme.com/health-traits/people-want-to-know-about-potential-health-risks/

8 https://www.nature.com/articles/s41380-017-0001-5

Notes to 6 Ethics and responsibility pages 143–157

1 https://calteches.library.caltech.edu/301/1/robert.pdf

2 https://calteches.library.caltech.edu/2532/1/theend.pdf

3 https://academic.oup.com/jlb/article/7/1/lsaa006/5841599

Notes to Conclusion: Genomics' place in the future pages 163–164

1 https://www.globalca.org/copy-of-about

2 https://doi.org/10.5334/bcq

Index